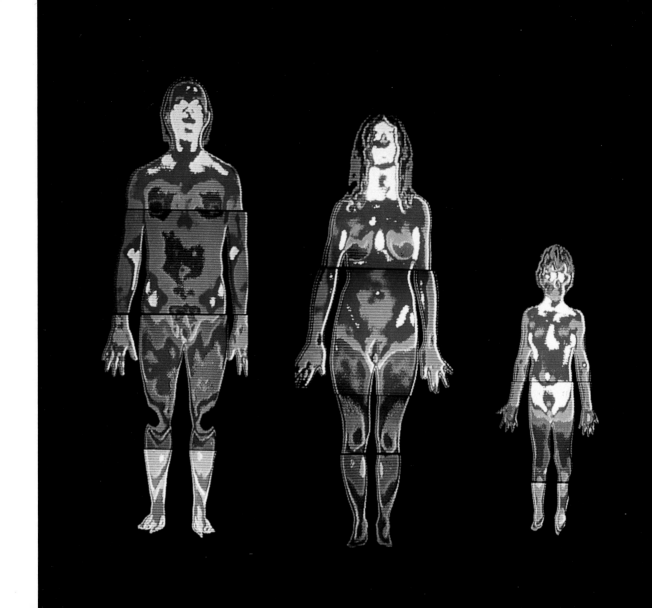

david bodanis

A Touchstone Book
Published by Simon & Schuster

the

secret

family

twenty-four hours inside the

mysterious world of our

minds and bodies

TOUCHSTONE
Rockefeller Center
1230 Avenue of the Americas
New York, NY 10020

First Touchstone Edition 1999

TOUCHSTONE and colophon are registered trademarks of Simon & Schuster Inc.

Designed by Karolina Harris

Manufactured in the United States of America

10   9   8   7   6   5   4   3   2   1

The Library of Congress has cataloged the Simon & Schuster edition as follows:
Bodanis, David.
  The secret family: twenty-four hours inside the mysterious world of our minds and
bodies / David Bodanis.
      p.   cm.
  Includes index.
  1. Human biology—Popular works.   2. Human ecology—Popular works.
I. Title.
QP38.B594   1997   97-14809   CIP
612—dc21

ISBN 0-684-81019-0
      0-684-84593-8 (Pbk)

In Blessed Memory, H. Samuel Bodanis (1901–1966)

In Bountiful Joy,
Samuel Adam Bodanis, born 1993
Sophie Beatrice Bodanis, born 1996

contents

INTRODUCTION AND ACKNOWLEDGMENTS   9

prologue: 4 am   15

morning

1   AT THE TABLE   19
2   BREAKFAST CONTINUES   59
3   AROUND THE HOUSE   95

afternoon

4   MALL AND LUNCH   135
5   SEPARATE MEANDERINGS   171

epilogue: late night at home   207

PHOTO CREDITS   212
INDEX   213

introduction and acknowledgments

I got the idea for this book, or at least the first hints of it, years and years ago when I was a little boy in Chicago. I was the youngest of a big family (five older sisters) and delighted at being surrounded by these great adult creatures, who breezed in with boyfriends or homework, with their shopping, and phone calls, and mysteriously wonderful school parties and dates to arrange. I also marveled at the way our family was a machine that somehow—through a mix of discussions, arguments, and agreements—managed to supply all the living requirements of its members: the blankets and sweaters, the words, food, shoes, books, milk deliveries, and all the rest.

By the end of each summer, though, we were bursting out of our house and often our previous year's friendships, and it was time to leave for the long drive to my grandmother's farm, packing in the suntan lotions, sandwiches, slacks and hair bands, boyfriends' phone numbers, writing paper, iced orange juice, books to read, and caramels, along with the six children and two parents. (Pets, though appreciated at home, were not allowed to ride.) At first I couldn't understand why navigation involved having a map spread out in the front seat: surely it would be simplest for the government to put up signs saying TO GRANDMA'S FARM at the appropriate highway intersections. I was even willing to have a few extra signs added, including references to my best school friends' families, so that they would be able to navigate too.

And then—was I eight?—it suddenly hit me that it wasn't just my family in a car, or my neighborhood's families, but thousands, millions, of Americans driving here and there on the highways, and that each one was part of a family that we would most likely never meet, but that still had to face almost all the supply and assembly and sorting operations my own family did.

It was a staggering thought. Were there really other little boys with crew cuts and white T-shirts who had big sisters that teased them? There would have to be an unbelievable number of white T-shirts produced in America to dress them all. And then, where did all those other sisters live? How could they function properly without me around? I tried to imagine how many newspapers and gym shoes and cereal bowls would have to be made to supply all the families: how many school principals, and milk delivery men, and teachers of the ethical principles it was so important to learn.

Time passed, and I continued to be fascinated by families. As I studied a little science I delighted in the image of humans making do not alone, but in family groupings, as we live surrounded by invisible chemicals, energies, and microbes. Later I worked in different countries, and noticed how varying families would cope with the great crashing waves of emotions or financial need or politics that slammed up against them, how events sluiced through in totally different fashions, depending on how the family was constructed. And then—nicest of all—I started my own family.

And then I knew it was time to write.

Most of the research for this book was conducted at the University of London's Senate House Library, the University of London Science Library, and Oxford's Radcliffe Science Library, though I also was helped by library staff and resources at the Science Reference Division of the British Library, as well as at University College London, the London School of Hygiene and Tropical Medicine, Imperial College, the London Business School, the Institute of Education at the University of London, London's Science Museum, and the Wellcome Trust.

In almost every area there were a great number of professional journals to help, and my PowerBook merrily beeped warning tunes from Nina Simone, to the delight and occasional astonishment of my fellow researchers, as the hours taking notes wore down its battery. The most useful journals and their generally unpaid review staffs deserve a special thanks, including: *Acarologia; Advances in Consumer Research; American Demographics; American Journal of Clinical Nutrition; American Journal of Pediatrics; American Psychologist; American Scientist; Annals of the New York Academy of Science; Annals of Occupational Hygiene; Annual Review of Microbiology; Annual Review of Nutrition; Annual Review of Psychology; Appetite, Archives of Environmental Health; Behavior Genetics; Behavioral Pharmacology; British Journal of Dermatology; British Journal of Social and Clinical Psychology; British Journal of Social Psychology; Building Services Journal; Bulletin of the American Meteorological Society; Bulletin of the Psychonomic Society; Chemical Senses; Clinical Allergy; Developmental Psychology; Ecological Sociobiology; Environmental Behavior; Environmental Entomology; Environmental Health Perspectives; Ethnology; Ethnological Sociobiology; Gut; Human Neurobiology; Human Relations; Indoor Environment; International Archives of Occupational and Environmental Health; International Biodeterioration; International Journal of Psychology; Journal of Advertising; Journal of Chemical Ecology; Journal of Community and Applied Social Psychology; Journal of Consumer Research; Journal of Experimental Social Psychology; Journal of Family Psychology; Journal of Genetic Psychology; Journal of Marketing Research; Journal of Marriage and the Family; Journal of Nonverbal Behavior; Journal of Nutritional Medicine; Journal of Occupational Psychology; Journal of Pediatrics; Journal of Personality and Social Psychology; Journal of Psychiatric Research; Journal of Psychology; Journal of the American Medical Association; Metabolism; Mycological Research; Nature; Neuroscience Biobehavior Reviews; New Scientist; Nutrition Research; Oecologia; Pediatrics; Personality and Social Psychology Bulletin; Proceedings of the Society of Experimental and Biological Medicine; Psychological Bulletin; Psychophysiology; Science; Scientific American; Semiotica; Sensory Processes; Social Psychology Quarterly; Transactions of the British Mycological Society;* and *Trends in Pharmacological Sciences.*

Not everything could be found in printed sources, and many people went out of their way to be helpful as my questions multiplied. I liked finding that Luther had worried about the imminent decline of marriage, 400 years before our current Op-Ed pieces about it; now I enjoyed finding out such information as what had been in those massive quantities of gum I'd chewed as a kid. Disgruntled chemists and engineers at several major consumer products companies were an especially useful source for a range of matters, though thanking them by name would not aid their professional advancement. Individuals and organizations who can be more easily thanked include: Dr. Peter Addyman, York Archaeological Trust; Professor R. M. Alexander, the authority on biomechanical optimization, Department of Pure and Applied Zoology, University of Leeds; Sue Cavill at the British Psychological Association; British Telecom; Dr. Ian Burgess, Medical Entomology Centre, Cambridge University; Professor B. A. Bridges, Medical Research Council Cell Mutation Unit, University of Sussex; Professor Bradbury, Department of Physiology, King's College London, for details on blood-brain interactions; Dr. J. Empson, Department of Psychology, University of Hull; Jan Pieter Emans, Ciba Resources Center, London; Dr. R. C. Garner, The Jack Birch Unit for Environmental Carcinogenesis, Department of Biology, University of York; Dr. Malcolm Green, Royal Brompton Hospital; Adrianne Hardman, Department of Physical Education, Sports Science and Recreation Management, Loughborough University; Judy Hildebrand, Institute of Family Therapy, London; Professor Stephen Holgate, Immunopharmacology Group, Southampton General Hospital; Dr. Lewis Smith of the MRC Toxicology Unit and Dr. Paul Harrison at the University of Leicester; Professor Andre McLean, Toxicology Laboratory, Department of Clinical Pharmacology, University College, London; Dr. Gene Mooney, London Foot Hospital; the University of Munich's Institute for Empirical Pedagogy and Psychology; Oxford Brookes University's Centre for Science of Food and Nutrition; Dr. Peter Rogers, Institute of Food Research, Reading; The Royal Mail; Professor Mick Rugg, cognitive development, St. Andrews University; Dr. John Sloboda, music and auditory comprehension, Keele University; Dr. Peter Stratton, Leeds Family Therapy and Research Centre, University of Leeds; Dr. John Timbrell, School of Pharmacy, University of London; Dr. Richard Wood, Imperial Cancer Research Fund, Clare Hall Labs; Professor Colin Walker, Department of Biochemistry and Physiology, University of White Knights, Reading; Dr. Saffron Whitehead, St. George's Hospital Medical School. Special thanks go to Tim Lobstein and Sue Dibb at the London Food Commission. When Her Majesty's Government insisted there was no possible danger from Bovine Spongiform Encephalitis—"Mad Cow Disease"—oh, did the

Commission have information proving the contrary, as well as leads for almost every other food topic conceivable. The Islington-based Women's Environmental Network, and Ann Link in particular, was also an especially friendly and informative organization.

From all these sources it would have been tempting to create an unwieldy slab of words. I must say I tried, but at Simon & Schuster, Bob Bender, ably assisted by Johanna Li, suggested, cajoled, or just skillfully snipped away with the editing pens to reduce that effect. Once they were done, Maria Massey was brought in to further sort out the pronoun forest, as well as unbraid the most tangled sentences. At an earlier stage, Becky Abrams was willing to pause in her writerly afternoons at the Old Parsonage to help with the initial idea for the book. Susannah Kennedy knows a great deal about anthropology, as well as most other subjects, though with her modesty she would probably demur. She helped me a lot: talking over ideas, pointing out the areas where not everyone would be quite as interested as the author; even once breaking an international flight to stop off at Heathrow for a much besought editing session. Throughout, the photographers, researchers, and scientists of Michael Marten's Science Photo Library did an excellent job of creating or collecting the pictures.

It would seem silly to write a book about families while holed up away from one's own. This is why Karen and I were willing to go to an IKEA superstore, even amidst the family-churning hordes of a Saturday morning, and get our nifty expandable dining table. You set it at one size (medium) for meals, and after, when everything's cleared, you tug it wider, and lay out your computer and notes and nonspill coffee cup—this is especially recommended with an investigative toddler about—and then, while everyone's busy with their day's work, you bang away at yours. The blessings of sharing this with Karen, my lithe and graceful wife, cannot be overestimated. At times when the stacks of notes got too large for the IKEA table, I retreated to a separate study, but even this was rarely in total isolation, as along with photos and Sam's collected bark sculptures, Sophie, now eight months, had developed a partiality for sitting under desks, gurgling at this and that, examining the world. Sometimes she'd look with hope at some distant—but reachable?!—intensely dangerous object, and in the process suggest text ideas for her thought-depleted dad. My own father was born before airplanes, the Somme, and long before television. Yet my floor-exploring daughter can easily have children who live in to the twenty-second century. Perhaps a snapshot of this odd epoch we're caught in—the last few years of the twentieth century—will be illuminating for her descendants to see.

This book was begun before Sam, our first child, was conceived. As it is finished, Sam is now three, old enough to walk me to that study, careful not to spill the coffee cup he's honored with carrying, and setting me to work with the cheerful injunction, "Be imaginative, Daddy!"

It's all that a man could wish.

prologue: 4am

THE LARGE DNA-CREATURE WITH TEN EYES AND FIFTY FINGERS DIVIDES INTO ITS COMPONENT PARTS AT NIGHT, EACH ISOLATED IN ITS DORMANCY AREA FOR REGENERATION . . .

the husband   The houseplant on the bureau is breathing down quietly over him as he undergoes the automatic eyeball cleansing, trachea widening, pulses of sugar-releasing hormones, and the bladder pressure changes we all experience at 4 A.M. His dinnertime food is slowly being digested, dozens of his brain cells are popping out of existence, and his abdominal muscle cells—valiantly used in the gym earlier—are slowly being rebuilt.

the wife   She is asleep beside him, with her brain busy in the mix of automatic data resorting and wish fulfillment known as dreaming, oblivious to the cosmic rays leaking in from the ceiling, which her DNA repair mechanisms are now actively defending her against inside her skin, ovaries, and other cells.

the daughter   Aged fourteen, and fervidly dreaming too, under an abundant acne lotion film, in response to recent distressing hormonal changes.

the son   Aged ten, and finally sleeping off the sugar blasts from the candy bars, popcorn, caramels, jelly beans, milk chocolate, and peanut-butter buttons consumed the previous day. The computer terminal on his homework table is turned off, but a central computer elsewhere remains actively awake for him, steadily listening to the Internet to collect his e-mail messages from distant friends.

the baby   An unexpected arrival, just ten months old. Half awake and gazing at the blue-and-red mobile slowly turning above; trying hard to concentrate, with brain cells hooking up in new configurations, there in the half-light from the hallway night-light; smiling at the sounds of the family dog wheezing away in the hall outside.

. . . UNTIL SLOWLY, AS THE HOURS GO ON, THE FAMILY COMPONENTS AWAKEN, TO UNITE GRADUALLY IN THE EATING CHAMBER DOWNSTAIRS.

morning

1

at the table

Yummy Yummy Yummy Yummy! says the father, reaching through the clutter of baby-food jars and serving spoons on the sunny kitchen counter, trying to find the right type of baby food this Saturday morning.

"Yummy," the mother chimes in distractedly from behind her newspaper, hoping for a few more moments with the Op-Ed page before she has to do anything too maternal and helpful right now.

"Mmm-my," coos the ten-month-old in his high chair, red plastic bib on and yellow drinking cup clutched in hand, his eyes brightly tracking the opened jar.

"In for daddy," calls the father, bringing forward the heaped spoon.

"Yep," murmurs the mother, turning the page.

WaaaaggggHHHHH! screams the baby, eyes wide, pushing the stronger new food aside and grabbing for his drinking cup's straw.

. . .

The baby puts his lips to the straw, and his tiny lungs pull upward, removing hordes of fast-flying molecules from inside the straw. Fragments of cold empty vacuum—tendrils of distant outer space—appear inside there. As a result, the entire weight of the earth's atmosphere, stretching miles above this house, slams down on the liquid, rebounding it vertically upward in the dark tunnel of the straw.

There's a click as the radio is switched on, and the baby reflexively turns to hear it. For a moment, by chance, the dial is holding on a foreign language station. The foreign sounds appear just a blur to the parents, but not to this baby. Each phrase is crystal clear. At this age even the most ordinary of human babies has a huge amount of extra circuits in its brain, and so the ability to register any of the several hundred distinct sounds in all of the world's languages—the clicks in Xhosa, trilled *r*'s in Spanish, growling *ch*'s in Gaelic. It's universal. Japanese babies can hear the *r* and *l* distinction their parents can't, while American babies can detect slight differences in Japanese, or in Hindi, that their parents are oblivious to. The skill only lasts a few months, so the baby traveler, not knowing which language island it's going to land on finally, concentrates now, struggling hard to grasp the voice sounds, and thereby keep the circuits that receive them solidly in place. Only when the dial's shifted away to a music station does everything change. Multitudes of brain circuits that would have stayed intact if reinforced start crumbling out of existence, they weaken, detach, are harvested by prowling housekeeping cells within the baby's brain, and then are dumped in the bloodstream for removal.

The baby looks at its dad for reassurance, and in a reflex almost impossible for a parent to avoid, the tiny muscles controlling the pupils in the dad's eyes suddenly tug wider. Males who don't have children rarely show this universal sign of interest. Women are different. Whether they've had children or not, they're almost certain to reflexively widen their pupils—it goes up an average of 3 percent in area—when a baby looks straight at them.

And then the dad goes ahead and wrecks it all by finally giving the baby that heaped spoon from the baby-food jar.

Some baby foods are fresh and wholesome, but many are simply the means for manufacturers to get rid of things they couldn't sell any other way. In the past it would have been harder to do this because people could tell if something terrible was added in. It would just sort of float around and you could look in and run screaming. But there's an excellent way to mask it all now. Insert a long polymer at the factory, one that swells when water is added, so much and so burstingly, that it stretches over the various added substances, making them impossible to distinguish from the food. As an extra bonus, the wondrously swelling polymer allows you to add in so much zero-cost water that it is often the main ingredient in baby foods.

One slight problem is that the polymer that does this swelling and masking tastes, to be honest, like wallpaper paste. But this shouldn't be too surprising, as it actually is the main component of wallpaper paste. To cover up the taste, tomato purée is often used as it is easily obtained, at conveniently low cost, from tomatoes that are too decrepit or just too bruised to be sold separately. The bright coloring also helps with the second problem, which is that the paste itself comes out a revolting gray when it's first mixed with the water.

To bulk it up, boiled and skimmed pigs' feet extract is often used, though in a pinch the scooped inner pith of discarded fruit can be added, too. Chalk is often added next. It tastes about as you'd expect (though it beats the wallpaper paste), but it is white enough to mask any of the gray gloop that shows through the tomato coloring. Vacuumed straight from a schoolroom eraser it would be too dusty to swallow. But with the polymers of the wallpaper paste, it mixes so smoothly with the water that is the main ingredient that it can be swallowed without a problem. Baby rice especially is color-masked this way, and it can be up to one-third straight, scooped-up white chalk.

The paste and water slurry now looks better, but it's still not a selling point to say on the label that chalk dust, pigs' feet, water, and paste are the main ingredients. Something more obviously enticing needs to be added to sell the product. Often that's meat taken from the animals we recognize as usual sources. But it's rarely taken from the *parts* of the animal we're used to. Cattle for example are largely fermentation chambers on legs and so have hundreds of pounds of mucus-lined digestive tubing inside. They need this to hold their gallons of bacteria and plant fibers in place until they are excreted. Such mucosal tubing would also be hard to sell if it were labeled, but baby food has often been exempt from any requirement to label the exact part of the animal its meat has come from.

Bowels accordingly are one of the more common meat sources used in

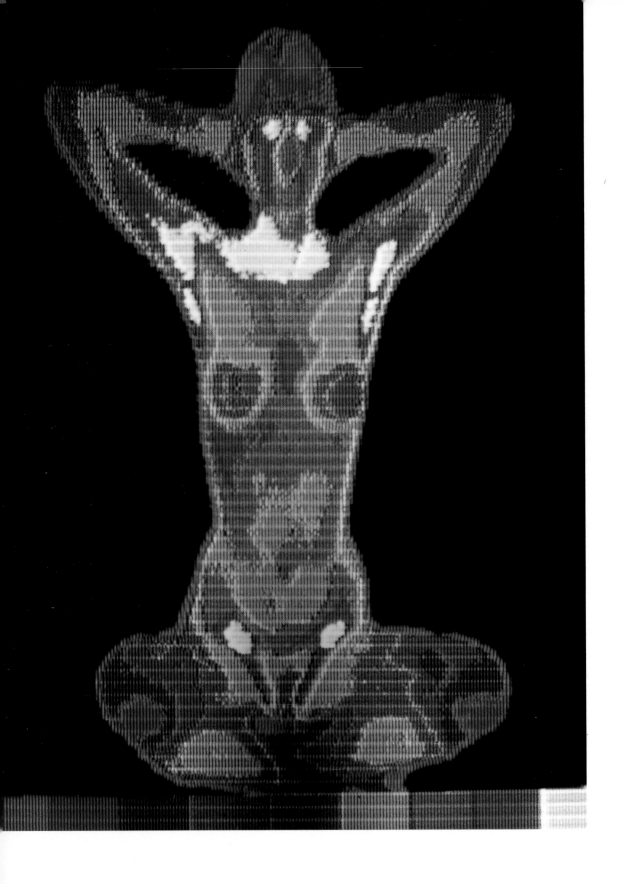

baby food. They are frequently put in great compression units with other hard-to-sell fragments—brains, testicles, and nostrils are especially common—and then they're all blended, squeezed, and cut into tiny cubes for mixing. If enough fat has been stripped off, the result can be labeled as "lean" meat. Sugar needs to be added to cover its taste. Kids wouldn't mind straight granulated sugar, but parents are fussy. Manufacturers, accordingly, often process fruit juices to yield a product that's chemically identical to ordinary sugar but usually can be heaped in without the dreaded s-word ever appearing on the ingredient list.

The mix is almost done, but it still lacks the right texture. The bowels, nostrils, etcetera, that came out of the compression units don't ooze with the connective fats that more normal meats would. Frothy chunks of animal fat get slopped in, to help along with vegetable fats as needed. A few herbs, an attractive label, some iron shavings to add mineral content, and there you are.

Usually that's it. But there are also some jars labeled as having extra ingredients which make them "suitable for the hungrier child." Sometimes it's processed cotton shavings or other cellulose pulp that goes in, other times it's just dollops of the dextrin glue used on the back of stamps. Both sound odd, but they're substances which swell exceptionally quickly once you mix them with water. Put the full water, chalk, bowels, pigs' feet, paste, sugar, fat, *and* stamp glue and cotton pulp mixture into a baby's mouth, and you can trust that he'll be left quite full.

The baby cries in horror at the next proffered spoon. The father is perplexed and tastes some of the spoonful, which seems pretty good to him. He only wants their tiny child, carrying all the parents' ancient DNA, to grow happily. But how can that happen if he doesn't eat?

Opposite: Thermogram of a seated woman. The hottest areas, seen in white, are the neck, ovaries, and armpits, where blood supplies are dense near the surface. Knees and nose are highlighted in the coolest blue; green is intermediate and often a sign of insulating fat layers.

The mother looks up from her paper, waking a little more—the increasingly sun-alert houseplants are spraying pints of oxygen into the room air, which helps. The mother and father consult. In a good marriage this is an easy task, for as they look at each other the parents are likely to see someone very much like themselves looking back.

With over 2 billion adults on the planet, the odds of any two pairing off are over 900 quadrillion to one. It seems an impossible task to ever get right, but there are a lot of characteristics we look for to make things easier. Some of the selection is purely physical. Eye color matches in married couples more than it would between random strangers, arm length matches and the length of ear lobes matches, and, roughly, attractiveness matches. It's only a rough match, partly because pretty women regularly end up marrying wealthy men, regardless of how the men look. (If he becomes less wealthy though, she quickly dumps him, as the statistics rather unromantically show.) But also the number of brothers and sisters you each had matches more closely than by chance, political opinions match, the amount of education your father had matches—even the likelihood of being a psychiatric case in future years matches.

About the only thing that's unlikely to match is the iron-loaded pigment called trichosiderin slipping out of the scalp in oil-lubricated fine tubes, producing what we know as red hair. People with red hair seem to hate marrying each other, and far more often than chance would suggest they pick someone with dark hair or brown hair or even no hair—anything to avoid someone with a red top. But it doesn't entirely work: red genes can be long-hidden recessives, and plenty of Americans with red hair have parents who aren't red-haired.

A final difference is more subtle. Spouses often smell differently from one another, but not simply because of the male reluctance to shower or bathe as frequently as women think that they should. Even clean humans generate a tremendous number of lightweight chemicals and these steadily float loose—a personalized invisible halo—into the surrounding air. Most are impossible to detect, existing just as a few isolated molecules, but a few hover more densely. These are the ones that enter the lower level of conscious awareness.

The effect is strongest on the immune system. It would be a great genetic advantage to select a mate whose immune system was different from your own, as that would give your kids an inherited boost through having more immune variations to build from. This in fact seems to be what happens. If women are given slightly used male shirts to sample, they almost always prefer those belonging to men with a different immune cloud from their own. When researchers go back and check, they find that the women's husbands or boyfriends also usually have a different vapor cloud. Which dating partners from the past were cruelly dumped simply because of the wrong immune system cloud?

A twist occurs when a woman has been taking the contraceptive pill. Probably because she can't sniff out other immune clouds as well then the tests show that she ends up feeling that the sexiest vapors are from men with *similar* immune clouds. Let her change contraceptive methods later though and trouble can arise, at least to the extent that we're chemical machines, for her vapor desires will take a swift 180-degree turn, and the same old spouse will be there. (Of course scent is only one of the factors that make up attraction.)

Over time, most of the main areas of spouse selection converge. The likelihood of drinking wine at the same rate comes to match in a married couple, the choice of breakfast cereal matches, and even the rate of going to art museums. People begin to resemble their spouses, and a couple are likely to go gray together, which is perhaps unsurprising as they were likely near the same age to begin with. The average is the man being three years older, a figure that holds in the United States, most of western Europe, and even in many hunting tribes.

The most curious change is in IQ. Here there's no bland convergence towards the middle. IQs match only roughly at first, but then—maybe it's those art museums you've been dragged to?—five or more years into the marriage, the score of the partner who had the lower IQ starts to rise.

Soon the older kids are called down. Only the ten-year-old boy gives signs of life at first; his fourteen-year-old sister is inert, still snuggled in her sheets, trying to delay the arrival of full consciousness. Early in a night dreams are brief, tetchy little things, twenty minutes long at most, and spaced at tedious, interminable, hour and a half intervals. In the morning though it's different, and if a teen can keep herself in bed and dozing, can squeeze her eyes closed and ignore the running dog and crying baby and her parents trying to locate the magazine section of the newspaper, attached electrodes would show her dreams popping into awareness almost continuously. Several neurotransmitters are sent out at lower than usual levels in the brain during dream sleep. Logical reasoning requires them to be present at high levels. That could explain the surge of pleasant—yet quite illogical—wish-fulfillment jumps we experience during dreams. Her body has been preparing her for waking—our blood volume and body temperature and adrenaline levels automatically start rising from about 6 A.M., in readiness for that—but with a little effort that tendency can be overridden.

The rush of dreams at the end of the sleep cycle, incidentally, is one reason jet travel from west to east—say, from California to New York—is so much more exhausting than the reverse. Since the eastbound travelers have to wake up at what seems an earlier time, they won't get as much time for those rich, world-orienting dreams at the end.

The boy emerges from his bedroom, and bangs with typical ten-year-old's delicacy on his sister's bedroom door, telling her that she doesn't have to bother about her beauty sleep, because it's not going to help anyway. She tries to ignore him, but her dreams are fading anyway.

They might deny it, but a brother and sister are more similar than anyone else in this home. The reason of course is that they have two parents in common, and all siblings share 50 percent of the same genes, on average. This is a feature they share with no one else. Everything from IQ to blood types to the immune vapor clouds and of course parental income levels is a close match.

More calls rise up the stairs, and although at first the girl shouts down with world-weary disgust that she's coming, the kids finally do descend, carpet electrons skidding underfoot to mark out their first house-crossing commuting trail of the day. The two kids bring computer games and colorful tracksuits and the joy of fresh life, yet also—quite regularly, in even the cleanest of well-off suburban homes—they bring something else, something quite small actually, cute to those who study it: an entire parallel family to the humans now assembling around the breakfast table. Actually it is more than a single family, it's family after family of little exploring creatures, strolling peaceably across the human foreheads, fond especially of the teenage girl's skin terrain, where they poke about for any leftover acne cream, or even just leftover mascara, which might make a nourishing meal. They can easily transfer to the boy and parents for further explorations.

These are the demodex mites. They're nothing like the awful, visible lice one can get from unclean conditions, for demodex thrive in clean homes. They're impossible to see with the naked eye, being only just the smallest dots even under a magnifying glass. But under an electron microscope they look like clomping mechanical cars, the adults each having eight pudgy little legs, on which they waddle slowly between their hair follicle homes. There's another, quite different, mite world we'll see later in this house, but the demodex are the most mobile and the only ones that make it to the kitchen. Virtually wherever researchers bother to examine, on our foreheads or near our eyes—a single eyelash will usually do—there the demodex are.

At first they can't really survive well out on the exposed outer reaches of the face. That's why the baby demodex, the cuddliest newborn pudgy ones, stay deep within the hair follicles on the face, nestling tight beside the soothingly warm root of a hair. Human hair can't grow without some liquid around its roots, and the baby demodex simply cascade upward on that, in a great sloshing water-slide ride. Most of us have at least a few bacteria for them to scavenge there. If the growing demodex has been lucky enough to land in the follicles of a teenager with acne, then there are supplies of nice, succulent, *propriabacterium acnes* and related bacteria waiting.

For about three days the demodex baby gorges and grows pudgier. Finally it changes form—an eager college graduate, putting on its best suit for the first day on the job—and then, for the first time, takes the final step forward, and it's out of the cave hatch and alone, tottering slightly, on the human face itself.

It's not quite isolated, for off in the hazy distance, there are other demodex graduates, also just emerging from their protective hair follicle bases. Amid the tentative newcomers are a few grizzled oldsters, cynical veterans of the tough life outside. The average demodex strolling across our faces is about ten days old. What it seeks—the stuff that's even better than the acne gunk inside the follicles—is all around. These are the fields of bacteria found on all human faces, generally quite harmless, especially the tremendously nutritious *pitorosporum ovale* sort. The walking demodex lower their pudgy heads and get to work eating.

A forehead-touching morning kiss is one way they spread within a family, but indirect methods are more common. The bathroom face and hand towel is ideal for moving from one family member to another.

In time though, as happens to all travelers, the lure of home becomes too strong; the isolated demodex grazers stop their feeding, lift their heads, and then, after a careful chemical sniffing of the environment, turn to go back. The ones still migrating, perplexed, on a bathroom towel are sunk, but the demodex safely on our faces have a good chance. It's usually about sixty hours since they left home, so the demodex that are heading back on Saturday morning will have been on the go since Wednesday afternoon. If there are too many eager youngsters at the original cave exit, or if it's blocked for some other reason, the pudgily returning demodex will have to find another home, but urgently now, for only at the edge of a follicle cave opening can they undergo their final growth, to become full adults.

The females stay at the hair opening once that happens, but the males don't have any time to waste and head straight back out. Now, for the second time in their lives, they have to wander over the landscape of the face. But this time they're looking for another follicle opening, one with a female, and they have to find it fast, for they're almost twelve days old now and don't have much longer to live. They trudge from one cave exit to another, repeatedly turned away by the other families already living there. Many don't make it and end up collapsing, alone, unmated, on the surface of the face, to be digested by the very bacteria fields they fed on, or possibly washed off onto a shower floor, if there's an especially thorough scrubbing by a hygiene-frantic human later this day.

The females have a longer life, and can wait up to three or four days at the hair follicle edge, for one of those exhausted waddling Romeos to appear on the far horizon and hurry closer through the bacterial fields, till there's contact, a discreet repositioning of bodies, and then the fertilization that will allow this particular genetic line to go on.

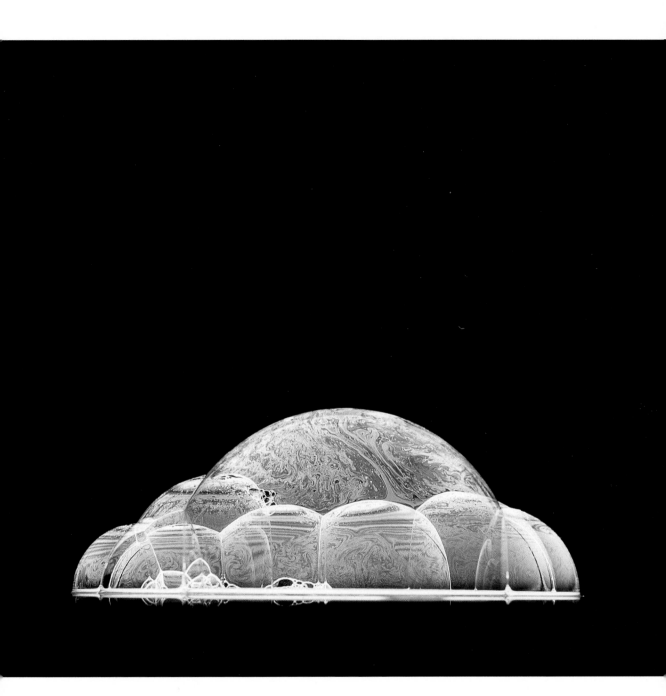

Wonders in a kitchen sink. At $^1/_{500}$ the thickness of a human hair, ordinary detergent bubbles are one of the thinnest substances visible to the naked eye.

Every human family gradually builds up its own distinctive population of walking demodex families. When the husband and wife first met they each had different subpopulations, from their own families. Only during the first few months, when they were dating or dining or tentatively living together, did those separate populations merge. Sometimes the result is true cross-breeding and a new unique group is created; sometimes, if one of the humans supports an especially hardy population, it's similar to someone who takes over a new relationship by insisting on carting in all their old books and old friends. The demodex newcomers out-eat, out-walk, and generally force away the previous ones on the spouse.

When their daughter goes off to college she'll carry a rich sample of the parents' now long-established population with her, and later, if not at her first Thanksgiving back then certainly by the next summer, she'll have fresh new-comers to offer, picked up from her foray into distant dorm rooms or rented apartments. But that collecting is only fair, as she'll also have left members of her parents' unique breed out there, hitching forever on the lush forehead and follicle worlds of roommates and boyfriends.

By now, waiting for breakfast, everyone's desperate for food: our brains run at a power rating of 20 watts, and this family's brains haven't been replenished all night. Murmurs, exasperated groans, and then the worst of all—whining demands from the kids—break the silence. The baby is lifted from its high chair, yelping at first at the floor passing so immensely far below as the great voyage starts. Its head tilts slightly to its right, which is the ideal reflex twist should the father continue the movement, to slide it into the much-sought docking position on the parent's left side. Here the adult's reassuring heartbeat is close below the surface. Babies are usually held there, and it doesn't matter whether the parent is left- or right-handed. Renaissance paintings and ancient vases alike almost always show babies being held on the left. The only klutzes are first-time dads who often miss the baby's suggestive head twist, but even they learn in time.

Still holding the baby, the dad opens the refrigerator with his free hand, releasing a blast of cooled air, which the baby sniffs. A family of two adults and three children will weigh approximately a quarter ton. In one year they will consume about 3,000 pounds of food and 900 gallons of drinks—this is why parents get that sickly grin when asked if they enjoy grocery shopping—so the refrigerator needs to be powerfully energized to store it all. Electrons are pulsing along the cable at its back, sliding forward a mere inch or so at a time, in response to the voltage charge hurtling through from the power station. If the station runs on coal, oil, or gas then this family is getting its food cooled by remnants of carboniferous forests and microscopic life-forms that waited 300 million years to yield up their final energy in each fragment of cooling gusts here. If it's a power station running on nuclear power, then an energy source older than our sun is being shunted through these cables into the house, for the uranium in these stations is simply the ash of earlier exploded stars.

Before such efficient cabling, chunks of ice were used to keep foods cool, but this wasn't easy to get. Mughal emperors of the sixteenth century tried, but when your headquarters are in Delhi this is a losing proposition. Riders started out from the Hindu Kush mountains, possibly with a certain amount of confidence if they hadn't done it before. But however well they packed away the ice chunks in their saddlebags, crossing the plains leading to Delhi the heat would bake and soak and generally blast into the saddlebags. In some miniature paintings from the time you can see the emperors and a favored few of their court with the result: tiny little ice sorbets. A somewhat more promising method was the New England system of carving up frozen ponds with horse-drawn iron saws, then coating the ice blocks in sawdust, and storing them in vast houses with double walls and insulated roofs. In the early 1800s, New York families who could afford it would sign contracts for delivery of 1,200 pounds of ice—eight pounds lugged freshly up their stairs each day—to get them through the summer. Ships could be similarly insulated, and by the mid-1800s New England ice was being shipped to the Bahamas and even—the Hindu Kush horse system long abandoned—to India.

Refrigerators profited from the ingenuity of James Harrison, an Australian printer. In 1851 he was cleaning blocks of metal type with ether and noticed how cold the metal got in his hand as the ether evaporated. He developed a machine where entire tubes of metal were cooled this way. The problem though, aside from the great size of the machine, was that the chemicals used in the coolant tubes tended to smell awful, and so long as that was the case home refrigerators had to be big devices, only for the rich, with the reservoirs for the tubings often kept in a next door room.

There was no solution to coolant problems, until finally, in the 1930s, one determined Du Pont chemist devised a mixture of chlorine and fluorine on a tough carbon core. The name it was granted accordingly was that of a chloroflurocarbon—he had just made the world's first CFC.

As the sounds of the family grow louder, the baby, having been put down on the counter, is enjoying its exploring too much to listen. Further along the counter a bowl of fresh apples are also talking among themselves: spraying out ethylene gas in simple data streams to coordinate their ripening sched- ule. (Fruit is often picked before it's ripe, and intentionally sprayed with this *"ripen quickly"* signaling gas when it's time to be moved to the stores.) The baby leans forward, and its moist breath cloud scatters aside the tiny ethyl- ene puffs, throwing the fruit's ripening suddenly off schedule.

The marauding baby turns back to his father, who's still looking inside the refrigerator, helped by light from the bulb inside. Inside that bulb is a filament glowing at several times the temperature of a blast furnace. This sends tungsten atoms furiously bubbling up, but the whole process is insu- lated so well within the oxygen-empty bulb that hardly any heat escapes. The ten year old is likely to have repeatedly flicked the door open and closed dur- ing the week to examine the light's mechanism, but although parents com- plain about imminent bulb burnout, there's little chance of that. The notion that lightbulbs use up extra energy each time they're turned on and off comes from confusing them with fluorescent tubes. Those do undergo a sparking jolt each time you fire them up, and it is more efficient to avoid turning them off if they're going to be needed again soon. Ordinary incandescent bulbs, though, in the hall or this fridge, are independent of flicking hands, just qui- etly evaporating until their filament narrows and decays enough to snap.

The baby can squeeze its eyeballs to get a sharper view of this illumi- nated box, but it still doesn't see what we do. The edges of the refrigerator and its parents faces and hairlines will be in sharp focus; it's the bits in be- tween, not yet fully comprehended because of its still incomplete brain, which remain slightly fuzzy.

The dad is having trouble finding what he wants. Men, on average, are exceptionally obtuse at processing information in three dimensions. Boys do worse than girls on tests about it at school, and often have problems even with simple inversion—That's probably why men are turning this page around while women are managing it this way just fine. The thick connecting cables of the corpus callosum that shuttle information between the left and right sides of the brain are usually less well developed in men; scans using nuclear magnetic resonance imaging suggest that a man faced with a three-dimensional packing problem is more likely than a woman to work primarily with just one side of his brain when he starts.

He's also likely to have a different search style. Women generally remember their way around paths on a map by memorizing particular landmarks. Men tend to skip those useful reminders and more often rely on pure guesses of relative distances and angles. In the refrigerator that might mean the wife knows to look for the fruit juice past the milk and behind the jelly; the husband is more likely to vaguely remember it as being up and on the left.

There also are strange seasonal variations, for higher testosterone levels sloshing inside exacerbate the problem. Men almost always do better at three-dimensional orienting in the springtime, when their testosterone levels are at an annual low. Men might be tempted to take some comfort in the fact that at least the total volume of their brains are greater than that of women—the difference is about 10 percent. But alas, such reasoning would only be a sign of these debilitating male limitations. Men generally weigh more than women, by that 10 percent or more, so of course their brains are bigger. A whale's brain is bigger still, but this does not make the aquatic creature a wiser beast. When you factor in women's smaller body sizes, their brain sizes are proportional to those of men, or indeed a little larger. Suffering all these handicaps, it's only with extra effort that the father finds the drink he was looking for and brings it back to the table to his waiting wife and kids.

HOW typical is this family? Despite America's recent high divorce rate, the most traditional arrangement remains common: about 75 percent of children still spend most of their time in a home with two parents. For a brief period after World War II the rates were even higher, but that was a historical exception. People didn't stay married for very long in previous times, not so much because of divorce but because adults died at young ages. Life expectancy in early colonial America was about thirty-five years, and in much of Europe it was less. Marriages were so short—in 1880s America they averaged just twelve years—that families were repeatedly made of half brothers and half sisters. The great number of stepparents in nineteenth-century fairy

A kitchen scouring pad. The yellow stripes are the nylon ribbons that keep it scraping effectively.

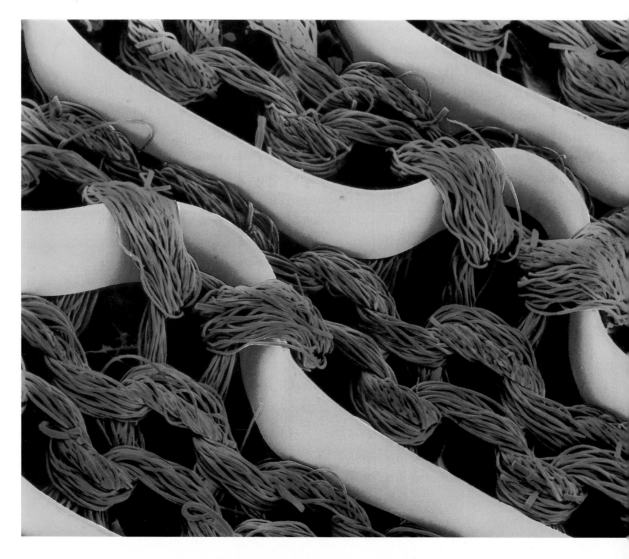

tales is a relic of this time, when almost everyone could count on a new family configuration at some point in their lives.

Right now though, every family member except the baby is sharing the first breakfast drink—a liquid which contains embalming fluid, varnish solvent, vinegar, and nail polish remover . . . and a certain amount of real orange juice, too.

The problem is entirely due to that last ingredient. Orange trees are so good at soaking up groundwater and transferring it to their little dangling fruits, that it costs a lot to ship an entire orange to market. It makes much more sense to squeeze the oranges near the place where they're harvested, and transport only the lighter-weight concentrate. When a local canner receives the shipment, it's accompanied by a manifest stating that, say, 12,000 gallons of water have been extracted. This is where the problems begin. An honest person would say yes, I must put back the 12,000 gallons and not a single drop more. The reason I can't add any more is that if I did, what I've received would go further, and be divided into more cans and jars, and that would give us easy excess profits, and that would be unjust.

Since not all canners are this saintly, extra water is often added—a few thousand gallons, often thinning it down by about 15 percent. But chemistry is an unforgiving science, especially when keen-eyed government inspectors are on your back. The water that's added back in can't be just any water. Natural water has a different isotope mix from what accumulates inside a fruit tree—inspectors can easily detect that—so the canner or bottler has to order what's politely called "pulpwash." This is what you get by taking the piles of tattered oranges that have already been thoroughly crushed and heaving them into waiting baths. Orange fragments scraped up from the surrounding cement floors can also be added. The mix is left to soak, the rind and pith will partially decay, and so a little more substance can be squeezed out later. That's the pulpwash. It's not really juice, but when shipped to the final canning plant it is added to the orange-colored cold soup. Certain popular brands contain up to one-third of this recycled pulpwash, even when they're labeled as "pure" or "freshly squeezed" orange juice.

Pulpwash on its own tastes terrible, so sugar has to be added, then some acids and a little acetone (the active chemical in nail polish remover) to give it some tang. When you do that though the acids start reacting with each other, so there has to be some vinegar to slow down that process, but you don't want to slow it down too much, so some ethyl acetate (varnish solvent) goes in. To keep it all from breaking apart, some of the embalming fluid formaldehyde—or a chemical near cousin—gets added as a final salvaging touch: the chemical is ideal at forming tight linking groups between proteins, be it dissolved bits of cadavers, or the more palatable pulpwash. Rail freight moguls who owned excess refrigerator cars from the Chicago stockyards began the fashion of delivering fresh fruit from warmer climes by advertising this concoction as orange concentrate.

The first sips of the refreshing orange, pulpwash, varnish solvent, polish remover, and vinegar mix go down, swallowed cautiously by the teenage girl, who's suspicious of any food substance that might contain calories, but rather less decorously by the ten-year-old boy, who's trying to glug in the way awesome beings—or at least the adults in those beer commercials on TV—should. Around the table each person's detoxification system immediately starts working: acetyl groups of atoms are prepared in everyone's liver and are added to each of the alien molecules as they come through. Without this process most of the food additives would get stuck, suctioned into our bodies' fat. With this molecular addition though most become soluble in water, so that they're easily led through the kidneys and on to the vast inland sea of the bladder for safe disposal later. It works fast: some of the quite toxic aluminum which the oranges pick up from the soil around their trees—aluminum is the third most abundant element in the earth's crust, so there's some spattered almost everywhere—gets fixed with this acetyl group so quickly that it's bobbing in the bladder of everyone around this table within twenty minutes of the first swallow. (Two organs get special protection against toxins loose in the bloodstream even before the acetyl groups get to work: one is the brain, and the other is the testicles. Unlike other organs, they are blockaded from all ordinary foods and blood travelers. They survive almost exclusively on a diet of pure sugar and pure oxygen, because very few other substances can pass their tight capillary barriers.)

A single chromosome location controls the main detoxification route, but it does not operate in every family at the same efficiency. In one unfortunate British company, twenty-three workers died of exposure to benzidine from the dye works they were tending. Almost all came from families who were especially slow acetylizers. The other workers had absorbed the same amount of benzidine but, possessing a luckier inheritance, had simply managed to float it away. In Asia a relatively high number of families are fast acetylizers and so are relatively safe from the estimated pound or more of completely indigestible colorants, stabilizers, preservatives, packaging gases, dyes, emulsifiers, anticaking agents, and miscellaneous metals a supermarket-rich diet will produce in a year. Most Europeans and Americans are only medium or even slow at the all-crucial acetylation, and so are exposed for minutes longer. One quirk which works in the opposite direction is that most westerners are quick to transform wine or beer when it first flows within their bodies. The reason could well be a genetic peculiarity that spread through the population. In Japan and China by contrast, more people are likely to lack the gene that ensures a fast breakdown. Sudden flushed cheeks—and a disorienting quick rush to the head—can result.

Now the mother doles out further health insurance: a vitamin pill to each family member, the guarantee that even if she hasn't been able to do all the shopping for fresh, nutrition-sopping vegetables she knows she should, she'll still be providing for her kids. But vitamins are actually tiny molecules, so the potions don't really have to be big as what's handed out now. Yet who would believe in the power of tiny pills? They're bulked up accordingly, and almost all of their volume is made of entirely useless fillers: there's sand and chalk and talc and, to bind the whole thing really tightly, steamed extracts from pigs' feet and other animal bones. Each ounce of actual vitamins the family swallows over time brings with it far more of the fillers. When the pill finally lands in the stomach, the vitamin at its center has so many of these extras to push through that the vitamin only slowly dissolves loose. About half of the most popular vitamin pills stay in such large granules that they can't be digested at all, and simply move along the digestive tract unchanged. The other brands are only marginally better, for unless a family is suffering from a severe food shortage most of the vitamins are immediately sieved out, leading, as pharmacologists put it, to Americans having some of the most expensive urine on the planet.

A century ago no family felt the need for vitamins, as food was thought of as one big undifferentiated thing, and all that counted was how much you got. This utopia ended with Casimir Funk, the Polish chemist who, with others, first recognized the importance of certain trace substances. He modestly resisted the suggestion of calling them Funkians, and instead selected the label of vital-amines, or vitamins for short. Had Funk been less modest, kids today would be encouraged to drink up their Funkian C, or take extra Funkian capsules to be sure.

The surface of these pills is often more potent than the molecules inside. Angina patients given placebos colored yellow rarely show much improvement, but when the placebos are red they produce a strong effect: enough to change blood pressure for days. Vitamin pills are often soaked in red dye to match such psychological (or placebo) effects.

The family discusses the next course, and also starts to plan the day's mall visit. A video would show a strange spectacle at this point, for even in the busiest of conversations we remain utterly still and silent about 35 percent of the time. The patterns of conversation are also odd, with most people repeating what they say in various forms, insistently, more than once, vocal chords whapping together, over and over again, until, finally, someone repeats it back to them and they stop. Little head nods from our listeners make most speakers go faster; if the listener is sitting with arms crossed most speakers go slower. The whole process usually begins with the speaker-to-be glancing away or down for an instant, as he or she seems to need this moment to plan the initial phrase. It's surprisingly hard to start talking if you're looking directly at someone.

What we see of each other while we're talking is strangely distorted, too. Everyone you're looking at first gets turned upside down and miniaturized, to fit on the inside of your retina. Then, as these images are shuttled along the optic nerve deeper into the brain, the truly surreal transformations begin. The family picture doesn't stay as a tiny reversed movie screen, but instead is broken apart into separate processing areas for hue, edges, movement, and the like. A vast number of isolated data bits, scattered throughout the multibillions of quick-firing wet brain cells, is all that remains of each person we briefly scan.

The son asks how much spending money he'll get for the trip, and the mother hesitates, trying to remember what they had agreed. What makes it hard for her is that an incorrect memory feels remarkably similar to a correct one. Dredging up either a true or false memory increases blood use in the part of the brain where stored data is accessed. The only difference is that when we access a true memory we're also likely to switch on the parts of our brain where the physical sensations of the true event took place. But short of bringing a huge brain scanning machine to the breakfast table to try to recapture that faint memory, how is a tired parent to be sure?

Letting the kids cajole, quiz, or even just speak around the table has not always been allowed. Erasmus, writing around 1500, was considered a dangerous innovator when he suggested children might be allowed to say something at meals "when an emergency arises." In wealthy Prussian homes in the late 1800s it was permissible for a father to punch a child with the back of his fist for speaking out of turn. This is not quite the American way, and tolerance for children expressing their views has exasperated European visitors at least since the Revolution. Some attitudes are surprisingly persistent. Children have been observed with their parents in playgrounds in Italy and in Germany. The Italian parents would hardly ever interrupt or hit their children over an hour's play; the German parents were going at them almost every minute.

Everyone's caught up in the family-linking discussion, except the red-plastic-bibbed prime member, isolated in his countertop exile. He's been watching just as eagerly as everyone else, but all the air bursts from the other mouths are hard to follow. His solution is easy enough. He concentrates, then creates a special nitrogen-vapor blast from his own mouth—the *scream*—to draw attention to the injustice.

The dad half fills the plastic cup with more juice, then passes it to his youngest. The baby smiles as he accepts this, taking it as his due, knowing he can now relax; triumphant, once again, in his essential parent-controlling skill. The vaguely orange mixture pours down his tiny throat; deep inside his body, the first quick detoxification starts biochemically whirring away.

And then, like a fool, the father asks his daughter if she'd like some more, too.

A creature furious with adolescent agony now emerges. Does the dad realize what he's done? she asks. Does he—does anyone in this whole family?—realize that she's been trying to keep to a diet, a sensible one, and not load up on this (a furious look at the proffered pitcher) this, sugary *junk* they insist on. Do they *want* her to not fit in any of her clothes? Maybe they've forgotten—hah, as if they'd remember—that it just so happens that she has a very important date coming up? No, she says bitterly, whipping open a magazine to flick through, already staring away from them all; they probably don't. They probably don't care. They . . .

The baby is enthralled, but everyone else knows it's best to look away for a while, and not argue back. The wife closes her eyes, and rubs her forehead. This is soothing not just because you no longer have to see your offspring, but because it helps change your brain waves: slipping from the jerkily uneven short pulses of ordinary attentiveness, to the smoother, more massagingly deep rhythms of alpha wave activity. Adolescence is not a new ailment. Aristotle recognized its symptoms 2,350 years ago, in the second volume of his *Rhetoric:* "They are hot-tempered and quick-tempered," he wrote of the young Athenians around him; they "can't bear being slighted, and are indignant if they imagine themselves unfairly treated"; they also "think they know everything and are always quite sure about it."

What makes it worse now—an exquisite prolongation of the agony for all concerned—is that it lasts so much longer. In Bach's choral accounts book of 1744, boys were listed as singing soprano at age eighteen. Girls also often reached puberty that late—even during the American Civil War puberty did not often occur until age sixteen or seventeen because of terrible nutrition. Poor countries today are similar. Among the Kikuyu of Kenya, for example, girls still only achieve menarche at age sixteen. The rich countries are the ones that have changed, with puberty beginning years before kids—these aspiring mini-adults—can leave home. Tension, and a drive for independence ensue.

You can make some predictions about how bad the battles are going to be, at least roughly. Families that talked a lot when a child was aged twelve have the best chance of ending up in that group of blissful households—57 percent of the total—where teenagers report actually enjoying their parents during adolescence. Girls who are especially pretty when they're eight or

Top: an immature egg, in an ovary where it's being nurtured.
Bottom: the fully matured egg tumbling loose. Mood- and body-affecting estrogens are released from the ovaries as well.

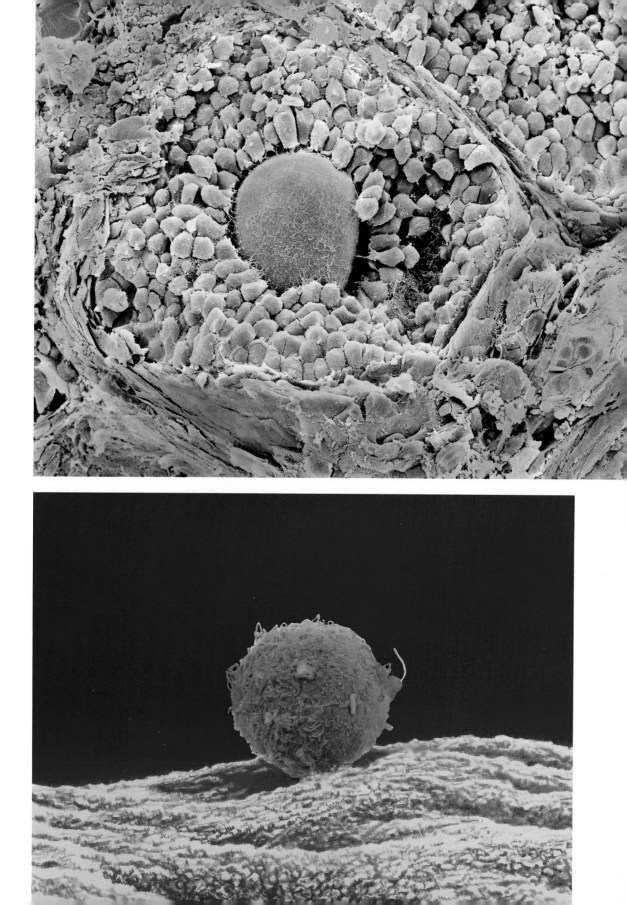

The chunky slabs of an ordinary eggshell. In a fertilized egg the slabs would steadily thin, giving calcium to the fetus for its skeleton and weakening the shell to make exit easier.

nine years old rarely end up in that lucky 57 percent, probably because they have more to lose from any change than their plainer schoolmates. And if a family's income has suddenly gone way up (or way down) when the daughter is starting adolescence, then she's especially likely to feel aggrieved, as she won't have a clear standard of family background to latch onto with her friends or boyfriends.

As the daughter flicks through her magazine, chlorine fumes from the bleached wood fibers billow unseen around her head. The father tries to return to his primary food-gathering role. But the hormone that's normally released from the brain to deal with stress, ACTH (adrenocorticotropic hormone), is likely to have overshot its mark, and several dozen nanograms will still be cascading down in a reverberating response to his daughter's flare-up. This can upset his immune system for hours, making him more susceptible to any cold viruses or other microbial assaults later in the day. The broken sleep because of getting up for the baby during the night makes it worse. When brain waves are measured in sleep labs, people with the most delayed responses from their immune system cells are usually the ones who've had the most interrupted deep-sleep brain stages. Only if the dad manages not to get too aggrieved—it helps to recite Ogden Nash's promise, that some day one's own offspring will get the delights of seeing their own children adolesce—might the peptide flow shut down quickly enough to avoid that need for imminent cold remedies.

A forced shrug of the shoulder, and the dad surveys the counter again, till he locates the yeast-pumped grain-dense sticky balloon known as bread. Over its surface an extruded polyethylene wrap has been plopping down tiny droplets of oily plasticizers and lubricants. Fifteen percent of a plastic wrap's volume can be such extra chemicals, put in to keep it smooth. In the family car parked on the street outside, the buildup of the day's heat makes enough of these smoothing chemicals evaporate so that they are detectable by sufficiently attuned noses later. Here most of the molecules simply pool up on the pitted grainy surface and slip off as the bread itself is pulled out. Fatty foods like cheese soak up even more. Most of the transferred molecules are harmless and easily detoxified, though some seem similar enough in shape to female estrogen hormones that if the wife ingested any amount when she was pregnant, the baby might have problems in his sperm production cells once he reaches adolescence.

White bread used to be the product of choice for the wealthy, as only rich people could afford the chlorine bleaching process and the sieving out of all the burlap, dirt, coarse hulls, and other extras with which peasants' bread traditionally came equipped. In France, for example, purchase of the fluffy white baguettes was a sign that a family had made it in the city or at least was not going to be tied to a farm any longer. Fashions changed when it was found that, although the burlap fibers and occasional powdered dirt could be left out with no harm, the dark hulls and other extras actually were better than the white stuff. Nutritionists regularly rediscover this fact, but does anyone listen? Even in 400 B.C. Hippocrates, the Greek physician, was haranguing his patients to eat whole wheat bread "for its salutary effect upon the bowel," but the records show Greeks of that generation and later still preferred white bread whenever they could afford it. Today's nutritionists also have to reckon with the stubborn resistance of bakers themselves, who generally try to discourage whole wheat bread sales. The oil in its wheat germ easily goes rancid, which means whole wheat loaves can't be stored as long as others. If you buy your bread very fresh though there's an unexpected boost. Most of the alcohol produced by the yeast that makes bread rise will have evaporated in the baking, but straight from the oven a certain percentage still remains.

The dad pulls out the first slices of bread, and deposits them in a shiny metal box, attractively glowing for the baby's interest. Inside, transformed starch molecules near the bread's surface are leaking out as dark ooze. This is the dextrin that makes toast easier to digest than unheated bread, as well as providing the crusty brown color. Toasting your bread at home would have been fatal to try in Cambodia when the Khmer Rouge took over, for all private cooking was outlawed. Every family had to eat at a central mess hall. If you tried to cook separately, and anyone saw you or informed, you would be killed.

Here though the dad is safely busy—in a world where parents have to work, isn't it fair to have fresh pancakes at least one day a week?—and he swoops the baby down to the floor. Eggs are brought out from the refrigerator. When he sees that they're running low he prints a note to get more and adds it to the crucial family communication device which architects repeatedly forget to supply: he sticks it up with a little magnet on the refrigerator door. Invisible curves of magnetic force swoop into the room's air, generated by quick spinning atoms inside that magnet. The iron atoms were created in a slow buildup over aeons in distant stars. The magnetic lines streaming through the dad have little effect on his body, but the ones contacting the metal door hold the paper on tight. Ancient Greeks thought that natural

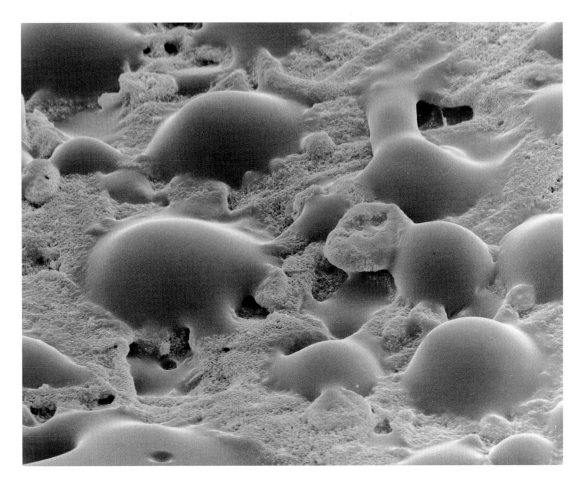

These igloos on the back of a yellow sticky note contain liquid glue: a few pop open each time the paper's pressed down, whence the repeated stick—at least till the last bubble's burst.

magnets came from an island entirely composed of the stuff, which anyone sailing near would be pulled toward and lost forever. Such an island has never been found—though who could report back if they got there?—and modern magnets are made by exposing appropriate materials to powerful magnetic fields supplied by a surrounding electric coil. The other portable memory holder on fridge doors—the ubiquitous little yellow sticker—almost didn't make it, for the 3M scientist who discovered their glue nearly discarded the formula when he realized what a poor adhesive it was. Only when he saw that this made the stickers uniquely useful for other tasks—marking his place in a church hymnbook was apparently their first use—did the promotions start.

The dad finishes his note and gets ready for the cooking. But which way should the egg be held? In *Gulliver's Travels,* two great kingdoms fought over the issue of whether it was the pointed or the widened end that was superior. Jonathan Swift, the author, thought he was being satirical, but this is simply because his comprehension of the buoyancy properties of the avian yolk was inadequate. When eggs are stored with their narrow tip down, then the air space inside will be in the wide open area under the spacious dome at the top. The yolk is barely buoyant in the dense medium of the egg white, and can't float upward. Only if some crude individual insists on storing the eggs with the pointy tip *up* will danger ensue. Then the air space is on the bottom, and the depth-bobbing yolk can easily bump against it. Oxidation or at least a partial drying out—not to mention a less attractive taste—will soon ensue. In any event, the egg's ovoid shape means it's unlikely to roll off the counter while it sits there waiting. Spherical eggs would easily roll away from a nest. Ovoid ones are safer, and in fact the eggs of birds that nest on cliff edges are often the furthest of all from a sphere, giving them the tightest rolling arc.

The shell splits open as the egg is cracked against the mixing bowl. The fresh yolk tumbles neatly over the mixing bowl edge. Buyers of free-range eggs might pause at this moment, musing on the little oval so recently in their palm; pleased that their free-range purchase, despite its whopping price, gave this health-bursting little thing and its parent a life, a real life, that was good and wholesome, and, well, *free,* there on its open farm.

It's a noble ideal, but not always quite true. For free-range eggs are merely covered by a law that says the chickens must have access to open territory. It doesn't say whether they have to be encouraged to actually go ahead and use it. Since it's no fun taking care of hundreds of chickens running around, manufacturers often try to make sure that the chickens stay inside. It's not really very hard; there's no need for little stubble-chinned rooster guards with Uzis. For chickens originally evolved in the tropics, and only if there's a space with trees and shade and running water or other Edenic accoutrements will they run outside. To keep chickens inside you just have to make sure none of those items are nearby. A tarred-over open surface with a chainlink fence around it is pretty good, and if a dog can be chained nearby or a narrow cement walkway built as the sole path leading from the main chicken roosts, you're doing even better. The "free-range" chickens can end up cooped inside just as much as ordinary factory-farmed birds.

The legal trickery is deft, but sometimes there's a way to tell when you've been cheated. Chickens that spend all their time inside often need artificial colorants added to their food to make up for their limited diet. The result is a giveaway unnaturally bright yellow color in the floating yolks. It's especially noticeable in the winter months, when even more food additives are needed. Embarrassingly lurid golden yolks can appear where an unpracticed farm manager overdoes the Technicolor dose. (Butter is often similarly colored, for although summer butter is naturally yellow, winter butter would naturally be pale. Here there are fewer giveaways though, because the yellow colorants in butter are easier to control.)

The chickens that grow inside the breeding houses are sorry-looking birds when they're finished. There are all the injections and strange feed products—recycled egg cartons and waste-drenched straw and even asphyxiated bodies of unselected males have been observed being ground up and used as feed—and because the chickens are bred to be scrawny, and are packed closely together, they're often pretty bruised once their lives are up. If only there were some way of getting rid of them, for cash, that wouldn't involve anyone from the general public glimpsing the shape they were in. It would have to be some very vulnerable creature that fed on these chickens, of course; one that had no choice about the matter, and might even be strapped down for the feeding ordeal.

The dad glances at the rows of yummy chicken baby food, stacked waiting to be used at the counter's far edge. But where is the ostensible consumer of those jars now?

The baby couldn't really be expected to stay still there beside his dad's legs on the floor where he was put down. Straight ahead there's the

The windblown shrubbery of an ordinary fungus, common on
bread or cheese, magnified 560 times.

partly open under-sink cupboard, and who knows what interesting things it holds. Usually there are only paper towels, but today he could be luckier, and someone could have put something away in the wrong place.

He crawls forward into the enticing dark, a miniature detection machine; eyes wide, and vapor-detection senses—babies on average have more taste buds and smell receptors than their parents—on the alert. He pads into a dense fog of radioactive radon gas that drifted up from the basement stairs overnight. This was stirred upward from the kitchen floor by the family's footsteps, but in here it just swirls around the baby's hands and knees. Some is swallowed. If the house were built on granite rocks there might be enough for lasting danger, but babies have powerful DNA repair systems—especially crucial with their active growth—so virtually all the damage will be repaired before the week is out.

Soaking into the baby's lungs are more substances swirling in the invisible fog. One is the common wood additive hexanal, trickling out from tiny pores in the plywood; another—especially if the family's been adventurous and done some repair work or put in new counters in the last few years—will be wisps of the universal stiffening agent formaldehyde. On a warm day, or when the heat has been kept on overnight, the levels down near the floor will be 100 percent higher than in the main house air. Some bounces harmlessly off the baby's clothes and hands; a little gets suctioned in from the baby's fast-breathing efforts to power itself, and ends up coating its lungs. But this too will soon be dissolved in the universal detoxification chamber known as the liver, as long as levels aren't too high. In a few hours some extra exhaled puffs of carbon dioxide—the formaldehyde's breakdown product—will be the only signs of its presence here.

The formaldehyde that slips past the baby explorer floats upward in the kitchen, but if there's a spider plant perched around, it doesn't stay floating for long. Spider plant shoots have a remarkable fondness for formaldehyde, especially when it is airborne. Microscopic holes open on the backs of their leaves, suctioning formaldehyde from the air. A few days from now, and the plant's roots will seem strangely energized, growing faster than ever, powered by this neatly captured aerial fertilizer.

The baby's in luck. Amid the harmless paper towels is a misplaced bottle of powerful cleaning fluid. These bottles have been as important as vaccines in fighting infectious disease and raising life expectancies: family mortality rates started dropping sharply in nineteenth-century cities, often decades before vaccines were available, wherever there was enough water and bleach and detergent.

Despite the waiting bottles, fungal cities are carving their way into the plastic and wood lining of the cupboard. Regular cleaning would get rid of them, but who bothers to scrub down here as often as he or she should? Fungi will carve their way into virtually any substance that has a toxicity less than that of uranium, which is why this cupboard, the refrigerator's rubber seal, the boy's unwashed socks, and, especially, that nice vulnerable bread lying exposed on the counter, are enticing sites for the digging parties. Kitchen fungi can produce up to one kilometer of fresh microtubing in a day. It's a strange cousinhood to have around us, for although we separated from fungi at an immense distance in evolutionary time, we still share many of the same genetic instructions. Fungal spores that have floated through the upper atmosphere to get to this kitchen—bobbing, unnoticed, beside the passenger windows of speeding jets—are likely to be sun-blackened by the operation of the same melanin chemical that is in our skin. Every kitchen is full of such uncanny parallels. The humble pea plant on the window ledge, for example, oozes out a chemical nearly identical to the hemoglobin in this family's blood, only in the plant its job is to hold oxygen deep in the potting soil.

Many of the arriving fungi can deftly steer around us in the kitchen air, using the flow lines of our airstreams to skim past human obstacles. The gust of the baby pulling open the door gave the fungi a good boost to soar to the food they crave. The furry spots we sometimes see on food are a giveaway sign that one of their colonies has safely taken root, and has grown an interlinked hydraulic feeding network that is such an engineering masterpiece on their tiny level that it extends up to be visible at ours.

The baby pauses to reconnoiter, lured by the pungent ammonia vapors from the misplaced bottle. Ammonia's ability to combine with water during cleaning also lets it combine efficiently with the micropools of water in the odor detectors within our nose. These ammonia vapors that lure the baby are the most abundant industrial chemical manufactured. The process used is remarkably subtle—the factories pull one of the raw ingredients needed directly from the earth's atmosphere—and it was perfected only under the pressures of World War I, by Fritz Haber in Germany, when a British blockade stopped the traditional ammonia-rich imports from South America that the German army needed to make explosives. Without Haber's invention, encapsulated in the humble cleaners today, Germany would have run out of explosives. Haber didn't get much pleasure from his invention: he was reviled by non-Germans as the man who prolonged the war, and even with the Germans he didn't fare too well. His parents had been Jewish, and as an old man the Nazis forced him into exile.

This baby-attracting chemical is also poisonous to drink. The family dog has far better smell detection faculties than any of the humans, as well, generally, as a fondness for the smallest member of the human family; as a result there's a bounding and face-licking and general infant-extracting ruckus as the dog rushes over to rescue its friend from the open cupboard. The baby giggles and tries to wipe his face; the wife shoos the awful, slobbering canine beast away. The dog keeps on wagging its tail though, delighted to be a part of things and stirring more of the low-lying radon in the process. Dogs can smell better than humans, not because their odor detection cells are built better than ours, but simply because they have far more of them scattered inside their wet noses.

Some people don't like dogs, but this is no doubt because they've been improperly introduced, or work in government tax offices. Anyone in a family can express his or her real feelings about someone else and get away with it through the simple expedient of looking the dog in its eyes and patting its fur and saying what he or she wants while pointedly *not* facing the family member who is really being addressed. There's also the pleasure of having this comprehending beast huffing and slobbering and nodding to you in deep commiseration during the soul-unburdening talk. Dogs are excellent for a family's health. The stimulus from patting dog fur seems to massage our brains. A good bout of fur petting brings together nerve signals that are rarely linked: there's the firing of the surface-touch receptors in our fingertips, and the warm-temperature detectors slightly deeper down, and then, in those moments of full-out they-don't-understand-me-either skin-pummeling rubs, the sort which domesticated canines put up with, as they're usually polite enough to give only the slightest squeaks of distress, the petting even gets our deep level bulbous-shaped pressure receptors, the ones normally untouched far beneath our fingertips, to fire their impulses brainward, too.

It works so well that regular petting of a dog almost always lowers your blood pressure; it even, as pleased researchers at the University of Pennsylvania have confirmed, raises the survival rate for heart attack patients. About 35 percent of the patients without a pet didn't survive for the length of the study; only 6 percent of the patients who had a pet died. Dogs can also be instructional. Mark Twain always gave his pets names such as Apollinaris or Zoroaster "to practice the children in large and difficult styles of pronunciation." Under the circumstances the enormous cost of dog food—it generally comes to 6 percent of an owner's total grocery bill—makes a little more sense. Despite all their merits, however, dogs are no longer Americans' most popular pet. With work demanding ever more time, during the mid-1980s cats took over.

It's true some cultures appreciate these tail-wagging health aids a little more than we consider entirely proper. In France there are twice as many dogs and cats as there are children, and wealthy women will often bring a dog to a restaurant and let it eat from the table. In parts of Polynesia dogs could have been brought to a restaurant also, but wealthy French women might have been upset by what happened next. Wherever a creature filled an island's niche of being one of the larger sources of mobile protein around, it made eminent ecological sense to use it for meat.

Up goes the baby to the counter once more, close to the intriguing radio/CD player and a safe distance from the pancake batter. The dad grabs the frying pan, puts in thick dabs of butter—for hasn't he read something distressing about margarine?—then adds a little extra water to the pancake mix. Dissolved lead might come out when the water faucet starts to run, having diffused in the pipes overnight from their soldered joints. If you let the faucet run for a few moments before filling any container, you'll avoid any such lead, but other chemicals are likely to come pouring out anyway, which is one cause of the Great Sperm Disappearance.

Something has been happening to men's sperm in this century, and nobody knows exactly why. College students of 1929 had 90 million sperm cells in each milliliter of semen, while their successors in 1979 could manage only 60 million, and today it's even lower. One likely cause stems from the way

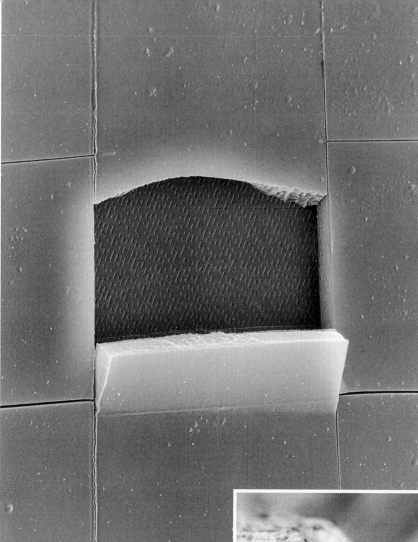

Two epochs of recording technology. *Top:* a compact disc, showing the digital notches under the protective plastic surface. *Bottom:* an old-fashioned long-playing record. The more the stylus wobbles through the plastic canyons, the louder the music.

farmers often feed or inject their cattle with female hormones to get the cattle to put on weight. These hormones are chemically near-identical to human ones, and they stick in food and run off into drinking water. At one point London's drinking water from the river Thames was so laced with female hormones that a cupful would give a positive reading on a sensitive pregnancy test. Other chemicals can also be part of the trouble—we encountered the hormone-resembling ones on the plastic wrap earlier. Expensive bottled water is not always an improvement, as labels declaring their products to be natural or pure or bottled at the source can still cover a multitude of origins. Sometimes bottled water will be what you expect, water from isolated rural aquifiers, untouched since the Stone Age; sometimes though it's just from ordinary wells, located near huge factories that pump in the natural-seeming carbon dioxide. In the case of one ingenious British supplier, the majestically pure water was produced by holding empty bottles under an ordinary faucet then sticking a glorious mountain scene on them. Whatever the source, this morning's drink of orange juice can only help: the vitamin C it contains aids sperm in its defenses against chemical attack.

The baby watches his dad in awe and even the ten year old looks up from his computer game. When there are naked flames, even the mildest of men can feel the primeval urge of the blacksmith pump through them.

It is Dad the Fire Wielder.

The Barbarian Swordsmith.

The squeezer of the teensy little button that ignites the safe gas burner.

The gas flame ignites, and molecules within the metal frying pan begin to vibrate faster from the rush of miniature explosions underneath. Some actually lift off into the food (though even if it's an aluminum pan there are not enough molecules, however long the cooking, to accumulate in the brain). Water droplets within the butter chunks start vibrating with the heat and themselves explode, and dioxin residues move up to the melting edge of the butter. Cows are the perfect dioxin-collection machines, chewing through great amounts of grass or grain that's constantly coated with fine traces of air-deposited dioxins. The butter pat bounces from the unleashed energy; in swirls the torrent of pancake batter, sending even more spattered fragments of butter flying up. At first it's the delectable butanedione liquid, which gives butter its flavor, but then harsher carbon granules soar up in their millions, floating like sharp-edged meteorites. The air fills with this abrasive rubble.

The heated butter could be dangerous, but the family's blink mechanisms simply switch to a higher gear. Normally we blink twenty-four times per minute, less if driving or happy or reading, more if angry or talking to a stranger. Children blink less than their parents, and cats, near catatonia at the best of times, make do with just two blinks a minute. The baby is closest to the line of meteorite impact so it responds first. Dangling blankets of skin weighing one-fiftieth of an ounce—the weight of a hummingbird's tongue—slither down over the vulnerable eyes. If our eyelids simply ground the microrocks into the eye surface the action wouldn't be especially useful. But with each blink extra salt water and flotation agents are injected from the tear ducts. The muscular skin blanket also curves itself in a gymnastic part-curl, not closing down all at once, like a fleshy guillotine crashing because of gravity, but rather rippling in sequence from the outside in, like a sea creature fluttering its tendrils in a sinuous wave. (You can just about see this directional twist if you are impolite enough to stand face-to-face with a fellow blinker and stare; you might as well, while you're at it, observe the other feature, that our more slender lower lid always rises first.) The push from the outside in rolls the landed rubble toward the small pink collection area at the edge of the eye nearest the nose. An exit tunnel there is quiveringly open, and leads down to the immense cavern of the nose below. The microboulders are shepherded to the pink area, a final blink squashes them through, and they'll only emerge several minutes later, free-falling or stickily dangling from the bottom of that tunnel, far from harm's way.

People tend to synchronize their blinking rates with each other, so the baby perched on the counter serves as an early warning system: the rest of the family, watching him from back at the table, is likely to start defensive blinking even before the aerial fragments reach the table.

High above, screwed tight to the ceiling, the electric circuits of the family's smoke detector begin trembling excitedly as the spatters float all the way up. There's a live radioactive chunk inside, which for years now has been fruitlessly beeping out alpha rays in order to create a smoke-sensitive aerial beam. Once such radioactive materials were dominant in open, intragalactic space and not humiliatingly consigned to life stuck inside a tiny plastic box. But now the radioactive chunk hints at its once-mighty power, as the alarm blasts away.

The dad is now entirely carried away with his fire-wielding force, and doesn't turn the flame down, so smoke rises into the enclosed room air. Under extreme conditions we can see the tiny burned butter granules rise as what seems to be a dark film of wispy smoke. The wife nods to her ten-year-old—a teenage daughter is long past the age of obedience—who steps to the sliding glass door and is about to open it to the back lawn.

The wife thinks the outside air will be entirely refreshing, but this is questionable and depends on what you mean by fresh. There are very old atoms waiting there, many having floated or stuck on our planet for several billion years, clustering up against the glass window. Each atom is a tiny speck shaped a little like a spinning mini solar system, tumbling weightlessly outside. The marvel is that we burn the stuff to breathe, but this in truth only applies to the waiting oxygen. There's the great bulk of seemingly useless nitrogen and other natural molecules. There are also, depending on location, all the manmade extras.

In Los Angeles what's humorously called fresh air is loaded, in addition to the well-known car fumes, with stupendous numbers of floating grease balls, small enough at their microscopic size to stay up for hours. They're generated by the estimated 3,000 tons of meat fried at fast food or other outlets in the area, and researchers have found that they compose 4 percent by weight of the fine-particle air pollution in the metropolitan area—a total of 46,000 pounds of flying burger bits per day. Blinking will push a lot of it across the eyes and down through the lachrymal duct to the nose, but a lot is breathed in as Los Angeles residents give their tracheal linings a burger-rich speckling.

The boy tugs and the sliding patio door is open, stretching wide. All the gases generated in the chemical warehouse of the home this morning—what had made the room bulge out balloonlike, as these particles struggled to get loose—finally have a chance to escape. There's a whooshing gush, as out it all pours, the house's unique pollution signature: the pancake particles; under sink hexanal; nitrogen compounds from the stove; and slow-bursting radon from the basement. There's even likely to be some CFCs slipping out from the fungus-weakened sealants at the back of the refrigerator. The smoke

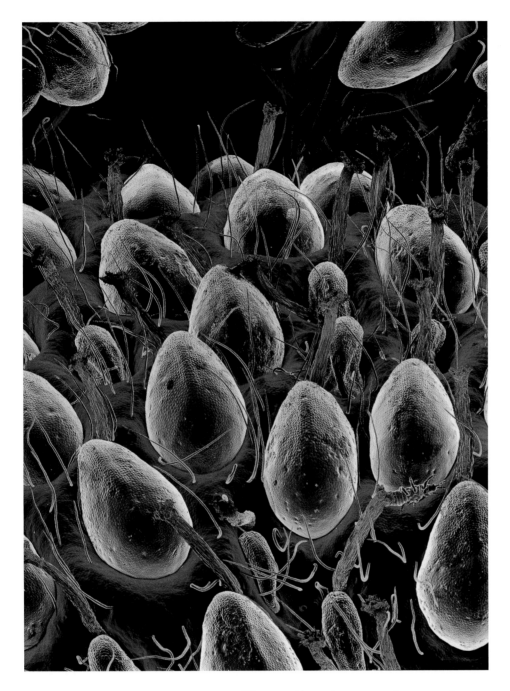

The swollen ripeness of a strawberry;
the green ovals are the seeds.

alarm quiets as everything speeds away across the patio and lawn. Some of the nitrogen compounds will travel entirely around the world in their twenty-year life spans before they decay. The CFCs will float up to the stratosphere, with the strongest-surviving of them bobbing in position for up to an estimated 460 years—a bequest from this morning's family to high-altitude explorers, if there are any, in the middle of the twenty-fifth century. But the family notices none of this, for everyone's at the table: the pancakes and maple syrup and toast with all its toppings, the orange juice and milk and fresh-brewed coffee, too, ready to be consumed.

2

breakfast continues

From the buzzer perched above the distant front door, a tidal wave of air suddenly begins to push into the house. A little bit races along the stairwell to the empty upstairs; the main wave pushes straight ahead, to the kitchen just across the front hall from the door. Some swirls over the fungal cities under the sink, but most of the crashing wave froths straight into the feeding family. Many of the high-speed shock waves crash against their bodies, bouncing off to eddy uselessly in the kitchen air; other fragments find the winding tunnels of the ear that lead deep inside each head.

And nothing at all happens.

The tiny metal clapper that banged into the bell pulls back and, still powered by the hand pressing on the doorbell outside, still a part of the first ring, now—about two milliseconds after the first strike—moves forward again and gives a second great resounding whack to the curved metal bell. Another tidal wave of air leaps out and races into the kitchen and hits the humans.

And still nothing happens.

The fiery sky of a sodium compound magnified. Sodium is omnipresent in the home, from a doorbell battery to the human nerve cells that hear the bell.

The metal clapper hits again, for the continuous buzzing ring which we hear is actually made up of numerous repeated slams, and there's a third house-filling air blast, a fourth, and a fifth—air waves blasting into the human ear tubes each time—and only at the 120th repetition, still a part of the first buzzing ring, a quarter of a second into the barrage and with the fungal cities thoroughly stirred up, the raised energy of the rushing air burrowing into chairs and curtains and fork-pierced raised pancake segments; only then does the first of the humans begin to react. For we live in an entirely different time scale from the mechanical objects around us; our reflex networks work through wet nerves and sopping brains, and we can't hear the separate strikes which made up such seemingly continuous sounds.

Slowly, with the gravest of deliberations, the father's eyes start rotating toward the sound source, sloshing in the lubricated eye sockets of his skull as six miniature muscle cables hooked into his socket bones start to pull. About twenty-five more aerial shock waves batter into the kitchen, and then, with their slightly slower reflexes, the wife and son begin the eye rotation, too. A moment later the baby's eyes start to respond, but by now the father, the comparative high-speed marvel, is gearing up to an actual body twist. Soon all the family's tendons are cranking and creaking, as they pull on their skeletons in order to get their heads and then their whole upper limbs to follow. In under half a second—a mere 250 repetitions of bell-thumped air waves from the first sound pounding in—the operation is done: all heads are tilted toward the source. Data streams of location, time, vehicle noise, and even door knocker banging can finally enter conscious awareness.

Although the father is likely to have started to turn first, the wife still has a chance of getting the hallway and the door beyond into view before him. The reason is that women have more receptors away from the center of their retinas than men do and so invariably register better on tests of peripheral vision. Men usually have to get a more face-on view to detect things as well. They also have to struggle in their responses to virtually every other low-level input. Almost every wife and daughter can out-see her husband or brother in low light. She can also usually hear high-frequency sounds better—such as any squeaking high-level harmonics from the buzzer, which is the cause of the dog's distressed barking—and, as we'll see, can usually detect dilute tastes better.

If there were a cat around it would do even better at peripheral vision. This is because cats have more rounded eyes than humans. These are great for bringing in the big chunks of light cats need for dusktime hunting, but the shape makes it difficult for the center of the retina to get a sharp focus— whence the curious way cats gaze into space with seeming mystic insight all the time. It's to make up for their focusing problem that they've developed enhanced peripheral sensors.

Ultrasonic shrieks, too high even for the dog to hear, are leaping out from the quartz crystal in the wristwatch, which the wife rotates upward to glance at. High-speed electrons lift off, racing upward from the luminescent coating on the dial surface, but these smash futilely against the thick glass dome, and stay trapped within. (Watches from before the 1960s used painted radium, which sent powerful gamma rays splintering through the dial case.) The habit of wearing chronological devices on the wrist largely dates from World War I, when it became stylish to copy Sopwith aviators who didn't have the time to fumble with an inside pocket. Dividing a circle into sixty degrees or minutes had been a standard practice since ancient Babylonian times.

After the turn, the next thing a startled family will likely do is get ready to swallow, for no one likes getting caught with his mouth full. Chewing speeds quickly rise from the stately molar descent of 0.08 miles per hour of ordinary times to a wind-rushing 0.12 mph now. Face muscles distort, and the motor nucleus in the brain controlling them is suddenly burning more glucose than before. Women have better control of the numerous tongue muscles and their direct lines to the brain than men, and so almost always finish chewing in fewer bites, as one can prove by surreptitiously counting. There's no reason to engage in the strange habit of chewing each mouthful thirty-two times. The practice, called Fletcherizing, after the turn of the century crank Dr. Fletcher, has no better justification than the fact that we all have thirty-two teeth. Children have more, for their future teeth are already in place in miniature form below their current ones, as X rays show. The tongue that pushes all the rapidly ground food into place reaches its maximum weight at age twenty-two, after which a small but steady decline begins.

The surprise at someone coming unannounced to the house on a Saturday morning is a sign of how much we've become used to being isolated with our families. Go back enough generations, and almost everywhere you'd have seen people rambling in all the time. One Nuremberg family of the mid-1500s reported, on a single Saturday, visits by peddlers, a jester, ice merchants, the priest, knife sharpeners, various servants' friends, a glass repairman, neighbors, a baker's apprentice, neighborhood children, and—the children's favorite—the leech carrier, offering hygienic or purely recreational bleedings. Hardly anyone in previous societies was granted much privacy—even Louis XIV was assumed to not mind his courtiers standing around as he urinated in the Versailles hallways—so there is no model for families being locked away all day. It's a bit extreme for our taste, but better than Mao's China. For many years most families there were allowed no visitors at all—not even relatives or close friends—unless the family had permission from a local block committee.

Once the hurried swallows are done, air is sucked in for a quick puzzled discussion about who it might be. The dad gets up to see, but even as he starts walking one inner part of the swallowing is still not completed. Liquids tumble in the free fall of gravity all the way down, but solid food goes slower, at the rate of the peristaltic waves of the esophageal tube pushing it along. There are about eight seconds—the rushed swallow long forgotten and the dad well on his way from the table—before the lower esophageal sphincter finishes widening, and, in a shared sequence of silent plops, four balls of the family's food join the earlier orange juice and toast, slapping into four waiting stomachs.

The baby bounces with arms raised skyward from his high chair, not wanting to be left out of the excitement, but the father isn't going to take him. The dog's efforts at bounding along are less thoroughly resisted though—who knows for sure what's out there? So one tail-thumping specimen of *Canis familiaris,* the rest of the family left protectively behind, accompanies Dad as he heads for the door now, pushing and wading through the kilograms of air.

There's barking and eager if not particularly accurate jumping, as the gush of *enemy?* vapors—undetected by the human, but frantically exciting to the canine—pour in from under the door. The dog is motioned away though, and the doorknob cautiously rotated. A grinding of metal parts now begins within the door mechanism as parts of the doorknob melt onto the rod that carries the turn to the latch. The pressure from an ordinary human forearm can do this because only scattered, isolated regions of the doorknob mechanism are melting—it happens only where fragments of metal are a little higher than everywhere else. But those scattered actions are crucial, for it's the friction they now generate that turns the latch. As you continue turning the doorknob, you quickly tear apart the first group of welded sections and replace it with another. (Polishing silver works on the same principle: for brief periods the point of contact between the cloth and the silver gets so hot that the silver flows over impurities and shines.)

A final shush to the dog, and the home owner steps out, to be blasted—this feels really good after you've been inside all morning—with energy from an exploding hydrogen bomb floating overhead. These arriving photons entered the upper atmosphere just fractions of a second before, when his hand was already off the metal knob; they were speeding through cold space in the orbit of Venus two minutes and twenty seconds ago, and only left their source—the surface of our Sun, of course—a brief six minutes earlier. Their history before that was more intricate, if a little slower: the originating photons began their journey up from the central depths of the Sun 10 million years ago, when dinosaurs were long gone, but *homo sapiens* were still in the future. A lot of these photons thunking down end up snugly caught in a cotton shirt or jeans, their 93-million-mile flight cut short just fractions of an inch from the skin below. But some make it farther, and reach bare cheeks or forearms.

Why do these fragments of a distant star feel so good? Their ultraviolet energy makes our endorphin levels rise pretty quickly. Endorphins resemble opium closely enough so that they trigger our brains' pleasure receptors, deliciously but safely echoing the effects of this drug. Other fragments of the light work on our eyes to trigger pathways leading not just to the usual optical processing centers of our brain but also to the pineal gland, where the melatonin chemical is made. In extreme cases sunlight can adjust the amount of melatonin leaking out and cure the depression some people get in the low light of winter; possibly in ordinary cases this has some satisfying effect, too.

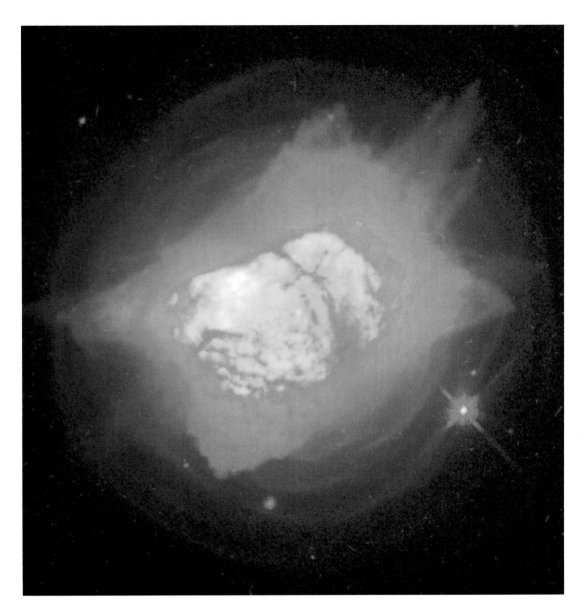

This star exploded 3,000 years ago, but its radiation is still reaching us, shattering DNA strands that our bodies must constantly repair.

Finally though, the light is also good for our health. Steroid molecules are packed in the top of our skin just waiting for the light. The impacting energy flicks those molecules into a new shape—what we know as vitamin D—and with that change they float off into the bloodstream, to reach the bones and other regions for healthful effects later. Stretch your arm out to get more of the rays, and your solar-powered vitamin factory switches to high.

Since the northern latitudes suffer from weather that ranges between miserable and very miserable, the effect is especially important there. That's why people who've had the misfortune to live for generations under those cloudy skies, such as the Scandinavians, get especially excited about sun exposure and come equipped with the most efficient reception units: pale, photon-transparent skin. Major suntanning is a different matter. The twentieth century began and ended with educated people fearing it. Only for fifty years or so, starting in the 1920s and inspired by Coco Chanel and other adventurous visitors to the south of France, was a contrary view widely held.

The dad stops to take a good sniff of the fresh air; proud of his manicured lawn and the shrubs and roses in front of the house. The house bricks behind him are a carryover from standards set centuries ago. They're just big enough for an ordinary-size mason to lift with his left hand: the right would be busy using a trowel. Some of the sweet vapor in the air is a desperate chemical message screaming out from the shrubs as they send an odoriferous alcohol upward to spotlight the point where they're being attacked by caterpillars or other gnawing insects. On a summer's day this SOS signal reaches any wasps lazily circling overhead, and leads them vectoring down to tear into the unsuspecting caterpillar. From the sensitive rosebush, another gas is being released to counter awful mass-fungus attacks. This gas leapfrogs over the fungus and signals the further reaches of the rose's leaves to pump food-blocking barriers around the fungus, starving them out. Even the attractive smell of freshly rained-on ground that we sometimes notice has a similarly desperate source: it's produced, in large part, by soil-dwelling bacteria spraying out poisonous blasts to keep away competitors. If someone's been too exuberant and planted the flowers too closely together, their roots are likely to start streaming hydrogen cyanide into the black soil to annihilate their neighbors, but occasional trickles will break through the surface, luckily at levels low enough not to disturb us.

The trees up above are a little more harmonious than the shrubs and flower beds. There's some evidence that they're hooked up by invisible gas communication lines, using puffs of ethylene gas to warn each other if one of them is coming under any caterpillar or fungal assault. The trees pump defensive chemicals into areas under attack, and where it's really bad, send out additional chemical puffs to help cut off a damaged leaf and send it, plus its bewildered caterpillar, circling in free-fall to the ground. Any caterpillar that doesn't survive the journey will be partially ingested by tiny tubes the tree sticks up from its topmost roots, in a sensible ecological cycling.

The dog bounds out, but the deliverer doesn't have to be scared: the creature's just racing to the trees on the lawn's edge, where the nose sniffer's heaven exists. Exuberant tail wagging helps this, by fanning odor communication molecules more briskly into the air. Any earthworm will detect the dog's shadow racing by, and will send danger chemicals back down the tunnels the worm emerged from, alerting other earthworms to stay down. You can usually trust that a new dog you meet in the neighborhood will be safe if its head is up or level with its body, if its tail is loose, and its ears relaxed. The time to climb a tree is when the head is bent down, the tail's taut, and the ears are erect. (If it's also crouching and staring at you, it's wise to climb quickly.)

To raise a dog that will let deliverers survive without blood transfusions it helps to have the dog handled by several different members of your family when it is very young—trainers have found this one of the best predictors of having a mild adult dog. It also helps if it's the right breed. Beagles and Labradors are on one side of the spectrum; on the other is the breed that was created in what might have been the plot of a particularly low-quality horror movie. Around 1900, a German dog catcher decided to save a select few of the dogs he collected from being killed. He chose the ones that had "the most vicious disposition, and the worst temper." The products of these experiments were crossbred with the next generation of most vicious strays. It took strenuous effort to make sure there was no backsliding over the generations, but with continued selection for vicious temper the procedure finally worked, and Herr Dobermann had the honor of getting a new breed named after him (although it lost the final *n* somewhere). In fairness to Doberman lovers, some branches of the modern breed have been crossed with terriers and greyhounds to make them less like their creator.

The deliverer's back at the sidewalk to check the house address against his clipboard. In Tokyo this would be especially difficult as many streets have no names. Each block may be named, for example, after a local shrine, but even so, the house numbers on it won't necessarily be in boring sequential order. Until about two centuries ago houses in Europe and America had no numbers. But when governments needed a better way to keep track of their citizens these useful tracking numbers appeared.

The deliverer climbs back up the steps now, the nearest ants turning curiously; the impact of his footsteps causing a great resounding thud on their incubation chambers below. He has a package in one hand, and now, in front of the dad, brings the clipboard up to be signed. But can this be done in just any posture? Social psychologists have noticed a shuffling almost as complex as the way ants frisk each other with their antenna: human males seem to have signed an agreement that they will never stand in direct proximity if they can avoid it. Instead, they take up a strangely distanced, now-don't-you-take-offense posture, with their chests not facing the other's directly. Such positioning would identify them as males even in the faintest sighting from down the road. Women, oddly unterrified, are generally willing to face each other directly, and stand closer than men will. They also give the little supportive nods and grunts we like from close friends during a conversation, and will regularly look at each other's eyes as they talk. Men, and especially men who don't know each other well, don't. If by some unfortunate chance their glances do cross, so that they're looking directly into each other's eyes, they almost joltingly flick their gazes away. The electrical conductivity of their skin goes up during conversation, and stops only when they finally separate.

It seems silly when replayed in slow motion on a video, but at least two individuals who know this code can usually get on all right. It's when different sexes meet that things can get confusing. Because men are so conditioned to sidle up from the side, thereby showing nonhostility, but women prefer to be approached cleanly from the front so they can see the other person, a man who approaches sideways in an effort to put a female deliverer at ease is likely to be met with suspicion.

The father has to sign for the package. The muscles that control his signature are simple to operate, but the inertia from the massive swinging hand sends the pen jerkily oscillating at the end of each stroke. This is why it's hard to forge someone's signature.

A pencil. Note the spongy wood—pencils are often made from trees 150 to 200 years old, to make sharpening easy. To make a pencil, the chunks of clay and graphite we call "lead" are placed in open halves of wood. The seams where the halves are sealed are often visible under the paint.

The dad can barely see what's going on around him now, for when we concentrate hard our visual processing ability fades away on the periphery. Our brain waves also change, briefly rising to a frantic cycling rate more than twice what they are normally. The blanking out of vision is especially strong in the first fifth of a second after taking in a new bit of information, such as a request to print one's name with each letter capitalized and in a separate box. Sounds are especially hard to take in then, and if someone speaks to you precisely at that moment, you will only dimly recognize the voice. Family members generally recognize this, and subconsciously adjust their speech rhythms so as not to interrupt during these brief phase-outs. Outsiders are not always so kind.

It should be no surprise which family name is most likely to go on the form: it's Smith. There are an awful lot of Smiths. It's the most common name in America, by far. The stands of every NBA basketball court could be filled exclusively with Smiths, you could have every team member be a Smith, employ as commentators and cameramen only people named Smith, have every actor in the commercials and every sideline cheerleader be selected from Smiths, and there would still be vast numbers of Smiths left at home to watch the whole thing. There are dozens of them at every big concert hall; the odds are better than even that there will be at least one in every large airplane overhead. There are over 2 million people named Smith in America according to Social Security estimates. With the equivalents in other countries—Lefevre in France, Kuznetzov in Russia, Kovacs in Hungary—the world total is probably around 5 million.

Many of the names on the clipboard—the Coopers and Bakers and Thatchers—make an interesting snapshot of the jobs that were common near the end of Europe's pre-industrial period. The reason is that this is when many family names were first consolidated, as with addresses, often for the same reasons. Yet not all the names that survive to be signed with a flourish today are quite so insightful. With stunning lack of originality, administrators in Greece and Ireland and Scotland all seemed to latch onto the habit of labeling as many people as possible with the names of their fathers. If that name were Michael, then Michaeledes or O'Michael or McMichael was the result. You can't do much late-medieval genealogy with that. And then there are the last names that might be illuminating but are still a little embarrassing for modern families to delve into. Kellogg sounds distinguished and hygienic enough, but similar names derive from "kill hog," commemorating a family of pig slaughterers. The common "ski" ending in Poland, apparently, comes from the period when most people there were little more than slaves. If you lived on the estate of say, a Count Pilsud, the easiest thing for the overseers to do at stocktaking time was give you, and everyone else slogging away there, the inventory label of Pilsudski. If the authentic Count also came to be known by that extended name, distant bearers of it could believe, or at least claim, that they were descendants of one of the elect.

• • •

The dog's ears lift up as it hears the sound of something familiar approaching around the corner. It starts barking in a different way, alerting the son, who knows what this means. He arrives outside just as a car pulls up, and his best friend, holding a white cardboard box, climbs out. Breeds such as Labradors hear sounds $1/8$ of a tone apart, which turns the generic rumble of a car into a crisp, unique chord. At one time the friend could have arrived on the fragile spinning gyroscopes known as the bicycle, but he's never been the leanest of individuals, and with distances these days, plus who knows about crime, that's gone. In England an estimated 30 percent of thirteen-year-olds have never been out of an adult's sight in their lives. In America, the figure's probably lower, but on the way up. The driver of the car, as chubby as the progeny he's dropped off, nods awkwardly to the dad on the porch, acknowledging that the driver should be embarrassed at off-loading his kid for the day without prior arrangement, but not so embarrassed that it stops him from gunning his engine and driving off to freedom.

Blood pressure and heart rates go up with all the excitement, but it's unlikely to have the same effect in the son as it has in his friend. About 20 percent of families carry a gene that makes their levels of the blood-clotting agent fibrinogen easily rise when they're excited. This pulls more cholesterol to the inner walls of their arteries. If the father and son have this gene they'll start to get a miniclump of oily cholesterol buildup now; the friend, even if overweight, is safe. Families with this common flawed gene will also get fibrinogen rises when the air suddenly gets cold, which is one reason shoveling snow in cold weather is so dangerous for a segment of the population. The worst possible time to shovel snow is Monday morning, for that's when heart attacks are at their weekly peak anyway. Saturdays are the safest time in the week, probably because general stress-hormone levels are down.

Exhaust fumes roll in from the accelerating car, but the humans on the porch don't get swamped, for the lawn's defenses get to work. The trees on the sidewalk's edge widen the open tunnels leading into their leaves, and miniature pumps inside start pulling in the car-released gases as they hurry by. Even a small oak can have 40,000 chemical-defending leaves dangling from its branches. The result is that dozens of pounds of poisonous gases and even the occasional toxic metal can be suctioned in each year, then shunted down to the hydraulic roaring roots to be deposited, as drops of poisonous liquid, far from trouble, deep below. The more defending trees you have, the greater the aerial scooping protection. The porch-gathered humans are now engaged in the footwork and mix of grunts of greeting and farewelling hand lifts needed to send off the deliverer and replace him with the visiting friend. Earthworms grinding through the soil absorb some of the deposited pollution, without too much damage; even tinier microlife in the stacked clay layers around the roots will get to work in the next hours and weeks crunchingly detoxifying much of the rest.

Air fumes that make it through the guarding tree barrier are next attacked by the lawn. Mists of invisible moisture are rising from the grass. This moisture coats the smallest pollution particles and makes them stick together and so suffer the worst possible fate for airborne travelers: a sudden overbalance of excess weight, and a frantic tumble down to the waiting grass below.

The *hungry* grass.

Tiny pores open wide on these miniature plants, too, pulling in the pollutants just as the trees do. Signals transmit from the grass's roots, alerting the soil's microlife that there's a food shipment coming down. But grass blades are so much smaller than the first line of defense at the trees and they're equipped with so much less shunting and detoxifying mechanisms, that at times individual plants are overloaded. Excess pollutants are dumped down to the roots, especially if the delivery van has been waiting, exhaust pouring, with a poorly tuned engine that's just spraying out the oxide clouds. It might look like the grass blades are just isolated towers of chlorophyll, each standing alone, but the lawn outside our house is an immensely busy ecosystem, with the seemingly isolated blades actually thickly cabled together underneath the surface. Leathery fungus arms stretch in a thick network connecting each part with others, and where one section gets damaged from absorbing the van gushing pollution, another section, safe from damage farther away, lush with excess nutrients, now starts to feed it. Attach a tiny microphone to one of those cables this morning and a miraculous gurgling sound would be heard as water and sugar and vitamins and amino acids all come pumping in.

Gutenberg's legacy: the corner of a computer chip, enlarged 60 times, active in appliances from modems to microwaves.

Back inside, the baby bouncing unhappily in his high chair, distressed at being left out, the chubby friend politely greets the boy's mother, then passes her—this is a regular ritual—the cardboard box. Everyone peeks inside, except for the teenage daughter, who is still ostentatiously sitting apart, but she wouldn't join in now if someone paid her. She finds it just unbelievable that they're willing to allow more food in this kitchen. Because they are Danish pastries though, everyone else agrees that the pancakes should be put, at least temporarily, on hold. The hand-delivered letter the dad has received is put aside on the table, too. The microwave door is opened, the cardboard box of pastries pushed inside, the timer is set, and the start button is pressed.

Is this wise? The theory is that all the microwaves are supposed to stay inside there, but crumbs or grease stains building up on the hinges make the seal less than perfect; a baby who's been eager for close-up investigation—with a sharply tapped spoon on the fragile protective grille over the door—will make it worse. A few of the high-speed microwaves almost always spurt out, squirming in five-inch-long shimmies as they explode into this family assembled raptly before its machine. Usually, regulators say, it's not enough for anyone to notice, but try telling that to Peter Backus, an astronomer with the international radio telescope in Parkes, Australia. When he first detected inexplicable signals at 2.45 megahertz on his viewing panel—and, even better, realized that they were coming at the same time every evening, precisely when one particularly suggestive stellar formation was appearing in the sky overhead—he was convinced he'd achieved the astronomer's Holy Grail, and detected alien life. Alas for the world's headline writers, Backus was a cautious sort and did some more localized checking first. As his shift started, the previous shift was ending. It always took a little while for his colleagues to get settled, which is why he usually had just enough time to switch on his radio telescope registers . . . and detect them heating up their frozen dinners in the staff microwave—set at the usual power rating, which is 2.45 megahertz—downstairs.

The friend, watching as intently as the family, is gushing tremendous numbers of alien bacteria into the kitchen's air. Humans pump-wheeze dozens of gallons of bacteria-rich breath clouds out each day, which is several quarts in just a brief wait. In the confined space of a breakfast room or kitchen everyone's going to get it. These bacteria are tough creatures that first evolved several billion years ago—almost at the first moment when the earth's surface temperature cooled enough for stable life—and have survived, truly dominating all other life forms on the planet, ever since.

What they didn't have to reckon with, however, was us.

Almost all these live microcreatures in the clouds the chubby friend is spewing out have a tremendously short life span here. The chlorine-rinsed sink counter and the bare metal stove top and the sleek plastic light switch are entirely inhospitable to their life requirements. They die by the thousands as they gently touch down, these parachutists from an ancient war. Only if you leave such nutrient-rich culprits as a slightly damp cloth or sponge around—both of which are dense in much-sought water, and with nourishing microfood fragments scattered within—will randomly landed bacteria find a safe refuge on which to grow. This is why most families end up with their own distinctive bacterial populations in their kitchens—and why eating over at a friend's can leave you with a low-grade upset stomach to which your hosts, after these years of exposure, are totally immune. Some people try to fix things up by using bleach or other disinfectants when they clean, but usually they think the right thing is to pour it down the sink. This is a manufacturer's dream—you get to sell something that people buy and then immediately pour down the drain so they have to buy more of it from you! Wiping bleach on the surfaces so that the invading bacteria have to try to survive a soaking in that is a better technique.

The wife breathes in, helping the baby turn over the envelope, and hundreds more bacteria are suction pulled entirely out of the room air, seemingly headed straight down to her lungs. That might seem better for the bacteria, yet the nice quivery lung tissue, all pink and vulnerable just a moment's flight inside, is not so easily reached. Everyone else in the kitchen is similarly breathing in whirlpooling bacterial emissaries from the newcomer. But a family that has defended itself against aluminum, frying pan rubble, and ultraviolet photons, is not going to be bothered by a measly fleet of ancient bacteria. The clouds of bacteria-rich air enter, but by the time the air reaches our lungs it is almost entirely sterile. Most of the bacteria are absorbed by the breathing tubes' sticky lining on the way down.

Other bacteria are carried toward the back of the throat, where with an unnoticed swallow or glug some two or three days hence, they will be heaved over the edge, to be sent down into the streaming acid rivulets of the stomach, from which there is no escape. We never notice when the sticky "escalators" carrying the debris up from the lungs works properly, and I suppose most of us would prefer not even to know of their existence. But for anyone with cystic fibrosis, where these sticky linings clog up, their proper operation is life itself.

Any ten-year-old who hasn't brushed his teeth recently is likely to have brought a rich additional store of bacteria, as an irregularly cleaned mouth is an ideal moisture- and food-laden incubation chamber for certain other species of bacteria. In particular, hunkering where the teeth erupt from the gums, there are likely to be awful spirochaete bacteria, with quick-spinning bodies, writhing like frantic miniworms in the hundreds of thousands all along tooth edges. Flossing would destroy them, for these are direct descendants from the earth's earliest life forms, and die when exposed to oxygen. Dental floss is a huge tugged rope on their size scale, and cracks open the oxygen-barring crust above them.

Even the spirochaetes are as nothing compared with one item that can be alive in the mouth of a boy who doesn't brush regularly. There will be enough micromeat to keep a tiny population of predators on the move. These are the *T. rexes* of the peaceful breakfast smiles: the awful *Entamoeba gingivalis*—a huge quivering thing, many times larger than the bacteria it hunts on the gums. A week or so of enforced flossing would kill the ones in place, but unfortunately that's unlikely to mean they're gone forever. An estimated 50 percent of household dogs have these same quivering monsters hunting on their teeth, and any eager bout of slobbery kissing by the boy will spread the *Entamoeba* right back. Chairman Mao *never* brushed his teeth, ac-

Beauty in a human mouth: undulating pressure ridges on a molar tooth.

cording to his personal doctor, but just rinsed his mouth occasionally with tea. Over the years his teeth became coated with a thick green film, no doubt rich with these microscopic hunters.

The boy notices his sister's aloofness, and with the typical sensitivity of a ten year old, glances to see his parents are occupied, then quickly makes a face and sticks out his tongue at her. Anyone watching with a quick-focusing microscope would now briefly get a glimpse of the great wobbling *Entamoeba* propelled through the air, straight toward her. But if she's lucky enough to harrumph in disdain at his immaturity, that will blow enough air currents to halt the invading monster before it can board, and send it, at its great weight, arching downward, to plummet harmlessly on the barren surface of the table.

While the rest of the family waits for the pastry to heat, the letter that's been delivered is distractedly opened, the perforated tab along its side scattering microclouds of cardboard rubble. Many will make it to the family's nasal air caverns, to join the squirming bacteria already stuck on the lining walkways. There's a bit of peering, even the youngest member stretching over to see, then a final deft flick of the tab, and the item inside is revealed.

Junk mail.

The parents groan and turn back to watching the microwave; even the baby, seeing their dismissal, tries to squeak out a matching groan in imitation. But to the ten-year-old boy this isn't junk mail: this is something addressed to the family c/o his dad, and that means them all, which includes him, and—who knows?—maybe it's important. The envelope announces a VALUABLE PRIZE to be won, so he and his friend start dragging their graphite-leaking pencils over the forms to fill out. The pencils' wood is liable to be much older than the boys, and probably older even than this house: to make pencils easy to sharpen, it's common to use wood from trees 150 to 200 years old.

The actual text and glossy pictures that fall from the envelope are left behind, ignored by everyone in the family. This is a shame, for the analyses worked out by direct mailers—the little electronic shadows of corporate information on us—can be so accurate as to surpass what a couple acknowledges about themselves. In many Protestant families, especially in the South, market analysts found that it's husbands who write the checks or sign the credit forms for buying the family car, despite what the women assert about being involved. The car mailings they receive, accordingly, show the men talking to the car salesman. Jewish couples turn out to be more likely to share car buying, so the car brochures that are mailed to Jewish neighborhoods or families with Jewish-sounding names have a better chance of including a paragraph on what fun it is to visit a showroom together. International variations also exist. In America, women usually purchase kitchen goods, however much their men try to proclaim that they care, they really do, about what goes on in the family kitchen. One big multinational has pictures of women thoughtfully choosing the appliances in the American mailings. But in the Netherlands, where the same brochure is used (with the text translated into Dutch), there are photographs of a man standing pensively before his appliances.

Everyone is analyzed. African-Americans are especially likely to receive letters offering encyclopedias, as from a history of efforts at self-improvement they buy more encyclopedias and educational reference books than virtually anyone else. Hispanics are likely to receive mailings offering patio furniture—even if they don't have a big yard—because the mailers know Hispanics feel obligated to spend above their income levels on such family-grouping items, however little they might actually want visits from all those relatives. And Chinese-Americans are unlikely to get insurance mailings mentioning old age and death, but instead they will more often receive solicitations emphasizing how insurance could help the next generation.

Often of course the mailings are misdirected or ignored, which produces an awe-inspiring amount of waste. An estimated half trillion items have been sent out during this century in America alone, with perhaps 500 pieces on average to each family in recent years. A suburb of 10,000 people would attract enough junk mail to build Noah's ark each month if all the paper were converted back into the wood from which it came; an American city of 1 million would attract enough to *fill* one ark every three days.

Rich people get more junk mail than anyone else, though here regional and ethnic differences break down. The effect is well known. ("You ever been black?" Larry Holmes, the champion boxer, asked. "I was once, when I was poor.") One sign you're at the top is getting packages with pictures of cars that look best with a château beside them, and what's on sale is the château, not the car. Another, and it's good to be prepared for these things, is getting unso-

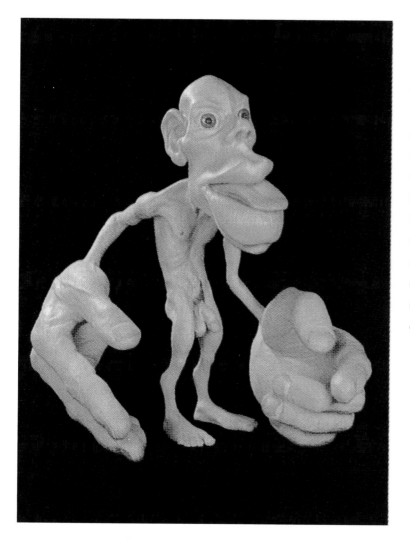

The lumbering shape of a sensory homunculus—a model that indicates how much space our brains give to the nerve signals cabling in from different parts of our bodies. Lips and fingers are loaded with nerve receptors, and so loom large; stomach and legs are more sparsely connected.

licited furniture promotions through your door that show pieces in forest green or burgundy. For some reason, these are the colors which the richest 3 percent of Americans are judged to select most. Do not, under any circumstances, feel complimented if you start getting furniture offerings in sky blue or grass green. The companies must know something from your financial records that you don't, for those are the colors which only the poorest 10 percent of Americans preferentially buy. And to be honest, if you're getting furniture promotions at all, you're not quite there. One British cabinet minister remarked with disgust upon a particular opponent of his, a mere millionaire, who was so deprived that he actually bought furniture. The slur would be lost on many Americans: a true aristocrat is someone who inherits all he needs.

The mailings can even tell you whether you're going up or down. "First-time buyers of investment devices" are likely to have been tracked by their types of cars or their neighborhood history; you and the other readers of that brochure are probably all going up. A mailing announcing "Consolidate those irksome bills through cash!" is a warning sign that the grass green furniture brochure is coming next.

The microwave buzzes, and the Danish—having, with a typical 2.45 MHz motor, been pummeled precisely 147 billion times in its one minute inside—is ready.

It looks delicious, icing glazed and caramelized, steaming with butter-rich vapors. Some people would deduce from this, that it's made of things like fresh icing and flour and butter. But that's no more likely than the family's fresh orange juice being made with fresh oranges. What we call Danish pastries—the Danes call them Viennese bread—couldn't be produced in mass quantities without some substitutions along the way.

The first thing to work on is that icing. It's hard to keep icing sugar enticingly white so the bakery company will just slap some white paint on. That's why dollops of titanium dioxide—the same chemical in the buckets of leftover white latex paint in the garage—form a good part of that gloopily delicious white substance on top. Where some brown, caramel-suggestive swirls are needed, brown waxes, including the indelible rosin used on violin bows, are often used.

Most of the rest of what's inside is pretty well known: flour, sugar, nuts, and oils. An ordinary bakery pecan roll can easily contain more fat than you'd get from a plate of eggs, bacon, and pancakes doused in margarine. But Danish pastry also needs to have what food psychologists (these are not researchers who interview foods—such people exist, but are generally only seen during visiting hours—but rather ones who ask others how they feel about foods) call "enhanced mouth-feel."

The simplest additive here, used in the most inexpensive Danish pastries, is made with dried extract of red algae, bleached so that no red color shows. It's cheap enough, just go to the right beach and there it is, and it does add a little of the desired stickiness. But this algae's quality isn't very high, and can also produce what's politely called "abdominal distention" if eaten in too great a quantity. Processed chicken feathers or the scraped belly stubble from scalded pig carcasses are often added to these lowest-price pastries, as their extracted proteins help in softening the flour that's used.

If you can afford a slightly better grade of Danish pastry there's likely to be a superior algal seaweed inside: the *chondrus* species. It's so much better and so useful for adding smoothness, that some is often used in cosmetics, which is something to think about when chewing a Danish while trying to put on mascara. But it's still nothing compared to the third and highest grade of mouth-feel enhancer. This is the high molecular weight polysaccharide called galactomannan. (Do not read any further if you like expensive Danish pastries.) Galactomannan is produced by processing the pods of a tree which originally grew in the Mediterranean region. It is an immensely sticky substance, ideal for mouth-feel needs. The tree, however, was also cultivated thousands of years ago. And who would have needed something so sticky in those years before it was included in Danish pastries?

When the Pharaohs of ancient Egypt died and had their brains pulled out through their noses and were embalmed, it was important that they be securely wrapped for their voyage into the land of death. This meant thick burial shrouds. What held those burial shrouds together, surviving so well, with a stickiness so undiminished by the centuries that you can still pull out the active molecules from just above the mummified bodies in museums today? It was galactomannan.

Most of the family can't taste any of these extras, and they eat contentedly: nostril skins twisting wide and rib cages convulsively expanding—our heart rate almost always briefly speeds up when we sniff food—to help whirl in the spurts of air needed for proper appreciation of these delicious vapors. About 20 percent of the population however are termed supertasters. Supertasters have more taste buds than other adults—about 9,000 is the average adult figure—and they even have more taste buds than the elevated numbers that children have. Very occasionally a man possesses this anomalous inheritance, but more often the sole supertaster in a family is going to be the wife.

(No one knows why women are better tasters. One idea is that it could be a way of detecting very low-level bacterial infections in food which wouldn't matter normally, but could be important during pregnancies.)

It's a lonely life. You end up spending certain mealtimes asking your husband, in whispers so as not to upset the children, if he's sure, really sure, he hasn't detected that smell you're sure is there. You will smell milk going sour a day or so before anyone else, and notice when someone left onions on the bread board a week ago. No one, ever, will believe you. This sensitivity is even stronger near a woman's monthly period, for the nose's inner lining thins slightly then, and incoming vapors are more acutely detected.

If the parents are distressed at their kids' eagerness in gorging on the Danish, they can take heart from an Ohio study done on children in the hospital. When the children were given their choice of any food to eat, they began as you'd expect, loading up on chocolate cake and other sweets for the first few days. But within a week—so long as there were no reprimands from the nurses—almost all of the children ended up willfully gobbling plain bread and fruit and—even the boys did this!—lots of chewy vegetables.

With the Danish eaten, the final harmony takes shape. The boys are immersed in their aliens-blasting computer game, and the girl has stopped bitterly flicking through her magazine—it's no good, really, if people have stopped paying attention to how dissatisfied you are—and is on the phone to a friend. She's most likely sitting down, for women almost always sit when they phone, whereas men stand, and she's probably holding the phone in her left hand, a curious family-linked trait which left- or right-handed people in a family are equally likely to follow. (One of the only sidedness traits more rigid in a family is which thumb you put on top when you sit with thumbs crossed over folded hands. It's passed on culturally, not genetically, but is usually 100 percent consistent among the kids within one family.)

She might be getting irritated by the distracting noise of her brother and his friend talking, but here there's an easy maneuver to help. If you switch the phone receiver from the left side to the right, you're likely to hear the words you want better. The reason is that the cabling leading in from the right ear goes preferentially to the parts of the brain on the left dealing with language processing. In a noisy setting, when someone is given two words at the same time, broadcast over headphones to his different ears, he usually registers the one coming in from the right best of all. This has some truly odd consequences. Tapping your forefinger is also controlled by cabling that stretches from the opposite side of the brain. This means that if you tap your right forefinger, then some of the language areas on the left are being used up. If you're asked to repeat back what someone says to you, then you won't do as well as if you were tapping your left forefinger instead. (Try it.)

The mother is finally able to return to the newspaper. Silent reading has not always been considered an acceptable way of passing time. A contemporary of St. Ambrose described the spectacle of first seeing someone reading to himself: "When Ambrose was reading, his eyes ran over the page . . . but his voice and tongue were silent. . . . We wondered if he read silently perhaps to protect himself." For almost all of history reading had been something done aloud. Early texts had none of the periods and commas and capital letters and paragraph divisions we're used to for helping out quick, silent scans. Only in the latter seventeenth century, with the spread of private Puritan contemplation, among other things, did Ambrose's strange habit become popular. We still have some holdover of the previous time in the few symbols some-

Eyeballs stretching forward from the brain, revealed in this horizontal scan through the head.

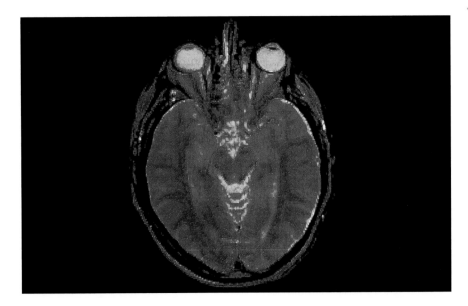

times put into our text to symbolize emotion and to help readers speak the words: the ancient "?" and the "!" Before the mandatory education laws of the 1800s hardly anyone was going to be doing any reading at all. In Spain in 1885 for example, only 500,000 people out of the population of 16 million could read and write. In Egypt and Saudi Arabia even today perhaps only 30 percent of adult women are educated enough to share the blissful morning relaxation of reading a newspaper.

The mother's eye pupils aren't moving smoothly over the words. Rather they hold still over an interesting chunk of text for about 200 milliseconds, then several of the miniature muscle cables digging into her eyeball give first a preliminary tensing and then heave. The eyeball starts spinning at impressive velocity to the right: so fast and so hard, that we find it impossible to see anything at all in these blurry rushing moments. After about 20 milliseconds, decelerating tugs from the other eyeball muscles start, and in another 10 milliseconds the blurry flight is over: the wife will start to process her visual signals again, with the eye neatly aligned with another chunk of text, further along the line. The only certainty is that when she comes to a period all heaves and tugs will cease: we almost always squeeze our eyelids and blink tight then—as again can be demonstrated by trying. Otherwise her blinking holds at a steady 18 or so each minute, though the more difficult an article, the fewer the blinks. All the while, oxygen molecules billions of years old land from the room air, to nourish the cornea sheet resting precariously just below her eye surface.

The dad glances over, and now yet another feature of eyeball movement comes into play. People don't just swell the pupils of their eyes when they look at babies. It's a far more universal sign of pleasure or distaste. When the newspaper page reveals a picture of a politician they like, their pupils briefly spurt wider; the facing photo, of one they dislike, will reverse the lens-rim muscles, and cause the eyelids to flinch. Because it's all unconscious, at times it can be embarrassingly revealing. If there's a young female movie star pictured on the page the husband's eyes are likely to quietly dilate; the wife's eyes will usually do the same for a male star she finds attractive. But men, the silly reflex machines, will usually dilate their pupils on just seeing the *word* "nude" in an accompanying headline, an excess women generally avoid. The sex differences go deeper. Men usually remember having seen a sexy ad but can rarely remember its content. (This is like Churchill's remark that he couldn't say he remembered Latin, but he definitely remembered having studied it.) Women, more sensible creatures, turn out to remember both ad and content, however much a hunk within it might have made them demurely dilate.

For both sexes almost everything that's read is going to be forgotten. An ordinary morning's newspaper can have 15,000 words or more, and if you add on all the office memos, backs of cereal packages, road signs, TV listings, mall ads, and even, for when we're really bored, the tiny print in the magazine ads: from all that our eyes might be faced with 100,000 words in a week. The daily *New York Times* alone, estimates have it, has more bits of information than the average seventeenth-century person would have come across in a lifetime. It adds up to 5 million fresh words each year; 50 million words—vast torrents of potential knowledge—each decade.

It's too much. The incoming letters get processed in the visual centers of the brain, broken into their component parts of angles and curves, sent swirling in quick, high-speed circuits to the advanced reasoning centers of our cortex, and then unceremoniously dumped. Only a very few signals, the ones that get further shunted through the region of the brain called the hippocampus, have a chance of becoming embedded in long-term storage. (After a stroke in this region people can end up living in a permanent present, utterly surprised at the presence of a visitor with whom they might have shaken hands and chatted just a minute before.)

The boys yell as an alien battle cruiser explodes, and the dad switches attention to the computer game they've been playing. The ten-year-old's brain is near its peak, pumping away with about 50 percent more oxygen fuel than the brains of the parents. The dad awkwardly tries to work the controls, but this isn't easy. University of California researchers have taken brain scans of people playing computer games. Beginners of whatever age go through desperate surges of neural glucose metabolism as they try to recognize what's going on. Experienced players don't, and can skillfully glide along with minimal brain exhaustion. Children even—this is what's especially unfair—are likely to have started growing different brains from their parents. Violinists end up with permanently enlarged clusters of cells in the cortical regions that control their left hands. (This is the hand that does the delicate fingering;

the right pretty much just saws away.) The younger they start, the more that region grows. This is a further reason parents suffer, while kids, with their analogous control-center enlargements, have to learn to be polite and wait.

The kids shouldn't be too cocky though. Out-of-date computers are often disparaged and the fact that Apollo 11 went to the moon with an on-board computer carrying only 16K of memory is regularly ridiculed in the computer press. But how many of today's young terrors with galactic blasters could program a spacecraft that would navigate to the moon with just 16K of memory?

Higher up, clouds of nitrogen dioxide left over from the earlier pancake cooking are dissolving harmlessly in the flecks of water within the moisture clouds the family breathes out, floating as invisible cumulo-nimbus formations over the breakfast table. Farther inside, within the nostrils them-selves, miniature rainfalls of sulfuric acid are taking place. This is because everyone's breathed in slight amounts of sulfur dioxide, which floated up

A group of X chromosomes, its fragile genetic information constantly being repaired. Individuals who lack a corner of the X in one chromosome have to get by with a smaller Y-shaped one—these misfits are called males.

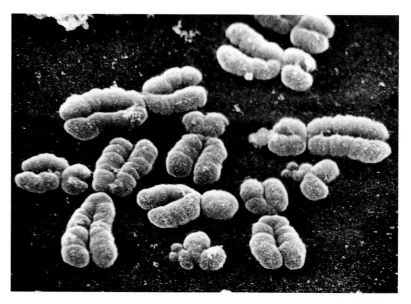

from the raisins in the breakfast cereal, where it was used as a preservative. Now it reacts with more of the moisture coming up from their lungs, to be harmlessly dispersed inside them. The kitchen's walls can themselves protect the family, for all the talking and plant watering and coffee-making lays down fine films of water on these walls. The abrasive nitrogen or sulfur compounds that land on the walls around the family are trapped. They bubble, dissolve, and then—twenty to sixty minutes after contact, depending on quantity—they're gone.

Detergent residues on everyone's clothes are another invisible menace. Skin defends by bringing up extra oil barriers, as well as replacement salts and amino acids as needed, to repair the local damage. Washing machines in America are generally worse than the best washers in Europe in this respect. The ideal washing machine is one that works with low suds—too many foamy suds and the sideways bashing that knocks the dirt out is cushioned—and has sturdy enough pumps to drain them all away. Machines that lack the drainage power, leave the skin-attacking residues. (When a wool cycle has been selected, an embarrassingly simple alternate mechanism operates. On most machines the wool cycle merely means that less water is poured in. The clothes are less soggy, weigh less, and so bang into each other with less thread-damaging whomps.)

Outside, the little patch of sunlight that's reached the patio is doing a good job of cleaning the air the family is going to breathe, traveling in through the open window. Ultraviolet rays kill thousands of live microbes floating in from out there. One reason we have fewer colds in summer months is that there are more daylight hours for ultraviolet rays to sterilize the air this way. The unobstructed sunlight also contains a fleet of tiny hydroxyl molecules that act as hovering garbage collectors, skimming along at window height and below, scooping up the methane leaking out from the family's plastic-lined garbage cans, as well as much of the carbon monoxide that's made it past the front lawn, and drifted around to this side of the house from the street.

Even the farthest traveled of all intruders are being quietly dealt with now. These are the cosmic rays slipping down from the ceiling, many launched from stars immensely farther away than our sun. A few are simply absorbed in the ceiling paint, seemingly exhausted after a flight of many thousands of years. The rest are still arriving fast enough to batter everyone's eye lens cells, which isn't too bad, as well as the crucial DNA molecules farther inside, which is. Without help your kids would soon look seriously strange, and the mutation rate would be too high for us to survive. Luckily, even as the wife's eyes jump to a new viewing angle on her mashed-wood page, repair molecules inside everyone around the table start snipping out the damage. Proofreading chemicals carry the broken fragments away, and while those are being replaced, the damaged bits are either reused, or are pumped toward the bladder, where they will end up further enriching the already aluminum-laden urine stored within. The outer space bombardment is held back a little by the earth's magnetic field, which stretches like an invisible arced dome over the home. Vast fleets of the incoming particles are constantly dragged away, out of the apparently empty sky, before they can get in and led to distant snowfields and ice near the Poles, where this magnetic field reaches the ground. Only in periods when the magnetic shield weakens—it seems to happen every few hundred thousand years—will the proofreading DNA repairs be overwhelmed.

The baby, meanwhile, is crawling along the hallway. He listens contentedly to all the sounds the adults don't hear: the air whistling over the polished front hall, the mail slot clacking, and the dog's breathing upstairs. There's a long moment's pause, then a final glance back to confirm no one is watching.

This journey will have to be fast, for a baby is still too insulated by fat to keep up high-speed movement for long without overheating. Its tiny 1.4-ounce heart begins to contract more powerfully, the eye-protecting blink rate goes up, and for the sake of keeping all its brain cells powered-up at an optimal seventy millivolts, its body sends spurting hormone signals to mobilize food sugars. Only then—with a final quick oxygen intake to make sure—does this portable exploration machine, little blanket gripped in one hand, set out.

It's slippery on the tiled floor, but saltwater squelches from the 27,000 or so sweat glands the baby has on each palm—a density three times higher than that of the adults so unadventurously staying behind. Thin ridges of skin—the fingerprints—sticking out on the bony extensions from those palms become hundreds of suction cups to help in giving a nonskid grip. In most families the direction of the fingerprint loops on the baby's and everyone else's middle finger will be aimed toward the little finger; in about 10 percent of families it will be consistently different and aimed toward the thumb. Finer details vary even within a given family, which is how the police measure of individual fingerprinting began. French police authorities were using it by the 1890s, though Scotland Yard is proud that a British investigator, Edward Henry, used them at least two decades earlier to control pension fraud in the army. For his efforts Mr. Henry received a baronetcy, and was popularly known as Mr. Fingertips. Chinese historians would be less than impressed: Middle Kingdom bureaucrats were using fingerprints over 1,000 years before that, as something like personalized credit cards for legal identification. There are similarly unique ridge patterns on toes, palms, and even on the roof of everyone's mouth, but those are less often left at the scene of a crime.

More cosmic rays are leaking down from the ceiling, but the infant's baby-size DNA repair systems activate. The tile's surface reflects bright light, which electric reception units in its eyes amplify and channel in for processing. The rush of colors and sensations could be confusing, but an exploring baby has been designed to do a lot of presorting. It won't worry about two light signals that are vibrating bare millionths of an inch apart, if they're both within the blue range. Yet two signals that close which happen to straddle the color boundaries its parents are teaching it—one perhaps on the very edge of blue; the other in the adjacent green realm—will be identified so the baby can concentrate on them. It peeks out the mail flap as fragments of Isaac Newton come floating up the stairs. This isn't because it's a house bought from a wild-eyed Stephen King–like real estate agent, giving the parents such a special price and answering with a haunting laugh when they asked why it was on the market at such a bargain. Sir Isaac is actually floating up the stairs of every family's house this morning, as always. The reason is that the human body contains at least 10^{25} nitrogen atoms, and long after a person's life has ended, a sufficient number of those molecules filter into the atmosphere to drift into almost every parcel of air. This baby's distant ancestors are rolling up with Sir Isaac too, finally to meet (and be breathed in by) their progeny. Believers in the transmigration of souls might take note. (There's also a certain amount of your own self always coming back, for every nine years or so almost every single molecule that makes you has gone, either

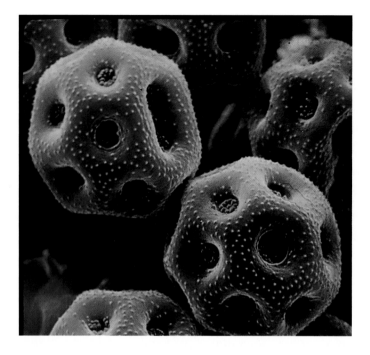

Pollen from a rose, 8,000 times life size, showing the pitted surface that blindly mates with flower or nose.

floated away or poured out. This solid stuff that was you doesn't stay entirely dispersed, and in its random travels some will steadily—in small parts—come rolling back home too.)

The baby looks up and glimpses the bright blue sky. It's blue because almost all the ancient photons which the dad was soaked with are absorbed by distant dust particles, and only a few—the ones moving with the frequency we call blue—rebound to reach our eyes. The baby stares at the sky-dome spectacle, blinking a little faster than usual as some of that otherwise invisible dust swirls in. Fragments $^2/_{1000}$ mm wide of distant forest fires, weathering mountains, and dust storms on the other side of the planet all land on his eyes, are squeezed over to the lachrymal duct pink areas and sent down for disposal in the nose and then stomach once again.

Bobbing in the morning air are live mating capsules, which have the habit of landing inside any watching human's nose, however young or small, and once in there, trying to mate. They quickly dissolve their uppermost surfaces, and dribble out enzymes. Then, still in the first few moments, they start to push forward their protein-rich hydraulic tubes.

Some people don't mind at all when pollen does this to them, and will venture out amidst the daisies and the hay fields, ready to enjoy a full, nose-lowered satisfying inhalation of air. Other people, however, would not leave their homes without tight-fitting gas masks on. Unfortunately both are likely to be in the same family, which is why shared decisions about summertime outings can be rough. To some extent there are genes that hay fever–suffering parents pass on, and which load their children with the reflex to make excessive amounts of the IgE antibody that triggers histamine explosions when pollen arrives. But not everyone in a family is equally likely to suffer. Oldest kids get hay fever four times as often as last borns. (It could be that they picked up fewer colds, as the family had fewer members when they were young, and that reduced their abilities to control any overeager IgE, but nobody knows for sure.)

Ozone gusting forward is more of a problem. The family members back inside are protected, for over the years their tracheal linings have become used to the usual levels of ozone floating up to this house. A baby hasn't had time to develop that protection. A lot sticks to his terry-cloth clothes. Ozone attaches a hundred times more readily to cotton than to glass or plastic and so is gushing directly faceward. The incoming ozone sinks into the sticky coatings on its breathing tubes, and there's an ever so faint tickling sensation as some of the cilia scaffolding underneath begins to collapse. In most parts of our body such slight tickling would be no problem, but the nasal mucosa always have to be protected. The sensitive endings of the trigeminal nerve pick up what the baby's receiving and send it through a direct conduit—there are special holes left open in the skull for it to fit through—to the brain. There's a brief moment of stillness, as quick cerebral evaluations take place; then the baby's face contorts in puzzlement, as the blast of sneezing air starts flying up. The vocal cords narrow, to force the air faster through the throat exit tunnel, and then, often with spectacular nasal-erupting force, out it all explodes.

Oooph! The baby plops down from the sneeze's recoil. Its small distance from the floor helps it land at a safely low acceleration, as does its relatively small internal volume. It's a general principle in the animal kingdom that smaller creatures are safer in a fall. As Gip Wells put it, a mouse can be dropped down a dry well and walk away, a human would get a broken leg, a horse would splatter. And there are advantages now, for from its sitting position, and nicely energized by the fresh air, the baby can contentedly explore some more, looking to see what's interesting here. The previous day's newspapers and junk mail left piled by the door are only worth a brief taste; far more interesting is that inviting electric socket beside the table where the phone books are kept. The baby lets go of its little blanket, then runs its thumb on the cool plastic surface, trying to push into the indentation where the powerful electric charges are connected, live, to the distant power generation station. Tiny droplets of water and gas are hovering in the air near the socket, attracted by the leaking electrical field. Because those droplets have radon-decay products stuck on them, the baby breathes in a quantity of radioactive products almost as great as what it took in under the sink. All electrical cables and gadgets seem to attract these radon-loaded invisible clouds: dishwashers, clothes washers, and bedside clocks are especially surrounded by them. Luckily the levels are high only if you lean right next to the cables or machinery, so the baby's natural defenses are sufficient. (It's possible that overhead power cables accumulate even more of these radon-loaded clouds, but the evidence is still being evaluated.)

Nothing much happens at the socket, so the baby turns to the interesting odor of his sister's leather-shiny black coat. Some of the smell is from the fragments of tobacco dropped from the cigarettes she secretly keeps zipped away in an inside pocket; more of it though is a long-lasting breakdown product of the nicotine she heats up when she smokes.

That nicotine doesn't last long—only a few minutes in the open air—but the cotinine chemical part of it is much more stable. Thousands of the cotinine molecules are tumbling down the sleeve of her coat even now, hours after she smoked with friends last night. They bounce in easy slow motion over her empty sleeve to float to the sniffing baby. Some will go into his bloodstream and end up, months from now, stored inside his growing hair. Even if the sister doesn't smoke much, but just hangs out around kids who do, some of this cotinine will end up there: the children of nonsmokers who are exposed to passive smoking almost always register some of it in their hair.

The baby reaches into the jacket, its fingers slipping along the shiny nylon inner lining. It's cool to touch because the vibrating molecules on his fingertips don't transmit their energy very well into nylon. That's also what makes nylon a good electrical insulator, and why it's likely to have been in the plastic wall socket. There's nylon all over the family's surroundings: door hinges, shoe heels, couch coverings, eyeglass frames, garden hoses, Velcro fasteners, combs, and car parts all are likely to have some of that same stuff in them. One might think that the man who invented such a ubiquitous substance would have been justly proud, but the scientist responsible, Wallace Hume Carothers, nylon's creator, was a tormented man who never believed he had done enough. In 1930 he created artificial rubber, but when this seemed to have no application, he started to work even harder.

In 1935 he came up with the polymer blend based on chains of six carbon atoms that became nylon. A number of name changes though, along with delays in getting a patent, increased Carothers's feelings that he was a failure. It didn't matter that nylon's ultimate commercial launch, on May 15, 1940, was the greatest success in Du Pont's history, with over 4 million pairs of nylon stockings sold in the first week. Nor did it matter that his artificial rubber discovery became crucial to American success in World War II, after Japan blocked rubber imports; nor even that, as one recent review noted, Carothers had created the fundamental understanding of polymers which now occupies the research of perhaps half the chemists in the world. Two years before nylon's commercial launch—a bare three weeks before its patent was finally filed—Carothers filled a glass with lemon juice, added potassium cyanide, and committed suicide by drinking it.

From a pocket of the girl's coat falls a small wood fiber box. This is interesting! The baby lifts the box, and little sticks with bright red tips tumble to the floor. It tries eating one, but the taste is not very good; it lifts up another one now and ponders.

At this point one investigating parent comes out, and sees the baby dangerously near the matches. The baby's picked up, away from this danger, and there are calls, angry ones, to the kitchen. The girl emerges, whispers hurriedly to her phone friend that she's got to go, then races up the stairs, sobbing. The boy and his friend glance up vaguely from their computer game, surprised at the suddenly empty table. They don't quite know why, but one thing is clear.

This breakfast is now over.

3

around the house

An hour later and the baby's happy, snug in his mother's lap, at her desk upstairs with the computer, sipping some nice warm milk from his blue plastic cup and watching her work.

Why is this so nice? There's the comfort of being held, and the murmuring of ancient counting rhymes—"eenie-meenie meinie-moe," meaning "one-two three-four," may have been carried over into English from the languages of pre-Roman Britain. But there are also the amino acids in the milk, including the one known as tryptophan. Normally only a little tryptophan will work its way into the brain, for there's no reason for it to be selected for cranial entry more than any of the other amino acids in milk. But if sugar or sweet chocolate has been mixed in, then the baby's pancreas squirts out extra insulin as it starts to drink. That helps hold back the majority of amino acids, and the tryptophan becomes dominant and has a free way up, up, through the pump-selection capillaries leading into the brain. Minutes after this skillfully dosed sugar and milk mix is in, it transforms into puddles of the powerful

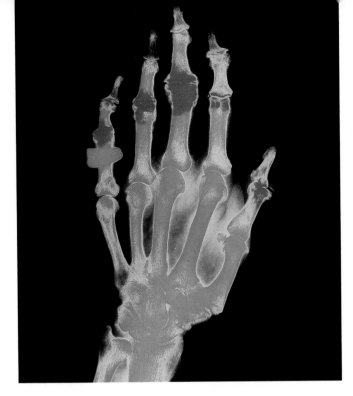

A woman's hand with the joints swollen by arthritis.

neurotransmitter serotonin. That acts like a miniature Prozac dose, leading to a smooth, tranquillity-inducing feeling. It works for adults as well as babies, which is why the old folk remedy of some honey in warm milk before bedtime makes sense.

A vast human hand reaches over the mellowed-out baby, stretching for its own midmorning sustenance; the wife's ring glinting in the computer-screen light. The gold of a wedding band is quite as old as the blizzard of immensely ancient oxygen molecules her hand is passing through. It was created in the supernova explosions of ancient stars, far more quickly than the iron atoms which had more slowly built up inside those stars, and fell toward earth as a fine glinting haze over 4.5 billion years ago. Any diamond on an adjacent engagement ring or, what's more likely till the mortgage is paid for, any preposterously overpriced tiny sliver of a diamond, is nowhere near as old as the star-flung gold or oxygen. Diamonds are crude newcomers, resident on our planet for a bare billion years or so; there are even some later arrivals, gate-crashers really, not even 900 million years old actually available in the shops. Many were created fairly near each other, and only moved upward in rough geological synchrony, in the subterranean super-plume episode of the mid-Cretaceous, when immense volumes of our planet's lower layers were slowly lifted skyward.

The finger which bears these much-traveled rocks is relatively shorter in women than men. Women are likely to have ring fingers about the same length as their index fingers; for men, or at least 70 percent of them, the ring fingers are longer, as an examination of the baby—the relative sizes hold from birth—is likely to show. The reason the fourth finger is selected for these adornments could be a holdover from ancient Egyptian beliefs that this finger, alone, had a favorite circuit through the body up to the heart.

The marriage which led to this consistently placed ring is not always seen the same way in different cultures. Polls show that Americans are the least cynical of nations here: when lists of possible reasons to be married are given—to have kids, for financial security—they insist it's none of those, but say marriage is good as an end in itself. It's the Europeans who are less impressed: in France and Germany today, large majorities say that married people end up no happier than unmarried ones. Britain seems to waver between the extremes. Charles Darwin, famously, made a less than romantic list of pros and cons while he was courting: marriage would mean having "to visit relatives . . . anxiety and responsibility," he wrote; but it would also, he noted in a parallel column, mean he'd gain "someone to play with—better than a dog anyhow." (Boldness won out, and with just seven months more procrastination, Charles popped the question, beginning what actually turned out to be a happy, decades-long marriage.)

It's no help this morning that few people can read a computer screen as well as they can read from paper. Our pupils constantly leap off course, distracted by random flickers on the screen, with the result that we go about 25 to 30 percent more slowly than on paper. A further problem is that reading from a computer screen exhaustingly dries out the eyes. It's common practice to sit straight ahead at a computer, which is odd, because no one reads a newspaper or book like that. We bend. Here before the screen, staring ahead makes the eyes bulge. More moist surface gets exposed to open air, which sucks up fluid from the eyeballs, and disperses it uselessly into the floating room air.

The wife would love it if she had more free time, though according to specialist instructors in time management she already does, but merely doesn't *recognize* that she does. One New York–based course inspired a female executive to gain extra time by no longer waiting, uselessly, passively, during that full minute each day she was stuck in the shower waiting for her hair conditioner to soak in. Instead she started using that time for flossing her teeth, to her immense satisfaction: "I've never had such healthy gums and glossy hair," she reported in delight. Another executive realized valuable moments during the flip-turn at the end of a swimming pool were being wasted. The solution here was to tape up waterproofed sheets of poetry, so poems could be memorized while swimming laps. Being pressed for time is not a development unique to the late-twentieth century. Only for the few decades before the mid-1970s were men's incomes generally high enough so that their wives could stay at home. Before that, women almost always had to work outside the home and fit everything else—kids, spouse, running the house, and maybe even fragments of time for themselves—into the brief intervals left over.

Like the baby, the mother is also drinking a cup of warm liquid. It's a concoction that she would be appalled to learn resembles mashed insect brains, chemically speaking. Early in the Carboniferous period, cockroaches and other primitive insects were already using two types of transmission chemicals in their brains: one, similar to our adrenaline compounds, would speed up the communication of signals in their bulge-eyed heads, and the other, similar to the baby's mellowing-out serotonin, would slow the signals down. For a long time no plant could artificially duplicate these neurotransmitters, but with the establishment of flowering shrubs about 70 million years ago that began to change. Many of the new species had cell mechanisms complex enough to build exact copies of the complex ring structure which insects and other grazing predators used. The result was the plant substances called the alkaloids, which include strychnine, morphine, and quinine. All are modified versions of insect nerve transmitters.

One of these alkaloids worked by being so similar to the original chemical which slowed down the insect's thought transmissions, that it would lock into that chemical's place in the brain of any creature that consumed it. The creature's own slowdown signals would no longer work, and the brain of any predatory insect that still insisted on seeking out the plant would go haywire. The plants which produced this alkaloid concentrated it, and dangled it down to be deposited just where the predators might come, the better to ward them off. It's the dangled concentrates—better known as coffee beans, of course— that we consume in the morning.

Coffee first became known in Europe during the seventeenth-century Turkish siege of Vienna, and spread so quickly that within a decade Pietro della Valle in Italy was complaining at how excessively "students who wish to read into the late hours are fond of it." By 1650 the first coffee houses reached England, and the oldest one is still there, located at Queen's Lane, Oxford (which is where these words are being typed, encouraged by the ancient plant extracts). Americans aren't as prolific consumers of coffee as the Scandinavians, who are world champions with an average yearly dose of 612 cups each, but they still manage to pour down about 33 million gallons of it each day, or the equivalent of thirty seconds' worth of what flows over Niagara Falls.

As the wife is occupied in perfecting her brain chemicals, the baby has sidled down to the floor, and is examining the curious little paper squares it has lifted from her desktop. It doesn't recognize them, and so, as it does with virtually all unknown objects at this age, pushes a few into the surefire chemical analysis unit of its mouth to get further information. There's definitely some sort of chemical coating on one side of the paper, quickly melting away and leaving a fun liquid feeling on its tongue. But with a further tongue-rubbed sampling there's now something reminiscent of this morning's baby food extruding from it, and at that the baby quickly makes a face and takes the square out; holding the wettened mass for a safer inspection in its hand.

Postage stamps are an intricate layered sandwich of chemicals. The glue is a true masterpiece of the chemist's art. Think of the problem. Not only does the glue have to be sticky enough to hold onto an envelope, but it has to be not *so* sticky that it grabs permanently on to your tongue. It has to do this with only the amount of saliva we're happy to dribble off, and then it has to stick to the envelope firmly, but still give you a moment or two to readjust its position. Finally, even once the chemists have worked out something that's tongue-attractive and humidity-resistant and briefly free-sliding, they still have to throw it out if it tastes bad. Or offends anybody's religion. Or is too expensive. Or too high in calories.

The solution is to mix baby food with Elmer's glue. The Elmer's—or similar petroleum-derived polyvinyls—gives it a sturdy hold, while the powdered plant starches of the baby food gives the concerned user time to correct any sideways-skidded placements. It's true that a little of the mix sticks to your tongue and unglues itself a few seconds or minutes later, so that you always end up swallowing a little bit. But that's no problem. The starch is easily digested, coming as it does from healthful corn in America or potatoes in Europe. The Elmer's admittedly is less nutritious, as it balls up into clumps of white glue in your esophagus and stomach, but at least those are microscopically tiny and in time dissolve. Dieters can be assured that even the most thorough licking of the back of a typical U.S. postage stamp will only give them 5.9 calories. Larger stamps can carry greater amounts, up to a gut-busting 14.8 calories, but convenient sponge-moisteners are usually available to avoid any between-meals temptations.

All stamp lickers, and not just thoroughly testing babies, pick up a further sequence of odd chemicals when they lick a stamp. These arise because stamps are normally stored in big sheets, and what's on the top of one migrates, at least in part, to the eventually to-be-licked gummy bottom of the stamp sheet above. There's usually a tiny amount of the embalming fluid formaldehyde, just as in the vapors from the under-sink cupboard downstairs and in the orange juice. In the kitchen it stiffened the bonds between orange pulp, or between wood or plastic grains; in the stamp it's put in to strengthen the soft, thin paper used. What seeps out is actually excellent at killing any bacteria that might be tempted to try to colonize the abundant starch fields of the glue; quality-control specialists will only have to add other bactericides for this if they find not enough formaldehyde is getting through. A little of the clay and chalk from the paper will have slid out onto the glue, too. Then there's some of the optical brighteners used in laundry detergents—on the stamp paper it adds brightness, too—and even some of the algae extracts

A guitar string, magnified 55 times. The core nylon strands are wrapped in a thin steel tubing.

used for moistness, as in the day's Danish pastry. All this floats up with each eager tongue swipe, to get swallowed with the corn granules and polyvinyl glue.

Back at the desk, the wife dawdles some more, taking another careful sip of coffee, then deciding she really should have a section of the grapefruit that's on her tray beside the coffee cup. This is good. The coffee she's been drinking may contain the strange chemical called kahweol, in addition to the caffeine mash, and kahweol, unfortunately, makes cholesterol levels go up in anyone who swallows it. Grapefruit will counter that, for grapefruit is rich in pectin, which binds away blood cholesterol, especially if one's bold enough to really dig into the pulpy membranes where the pectin is densest, ignoring, for the true pith scrapers, those troublesome spatterings on distant window-panes. Individuals who can't bear grapefruit, and who fear those awful pithy bits, can get around the problem by drinking only filtered coffee. The kahweol molecules stick to filter paper and are left harmlessly behind. Both are solutions that any cholesterol-anxious dad downstairs could have considered.

Her delays are tormenting the primitive silicon brain that waits inside her computer, and has been programmed to always, however much it's re-buffed, try to keep in contact with the World Outside. If it has an alarm function set for 7:00 A.M. next Monday, then at every second this weekend its program will consult the onboard clock to see what time it is, and if it's not yet 7:00 A.M. on Monday, which it won't be, not for another 20,000 seconds or more, it will accept this disappointment, calmly waiting, till its computerized instructions build up, and then, ever hopeful, it goes ahead, having totally forgotten what it's just learned, and checks again.

Unaware of these waiting shepherds, the wife lifts an oblong plastic box to her ear.

She'll waste a little more time by phoning a friend.

Most people think the ringing sound they hear after dialing is produced by the phone they've contacted, but that isn't so. The small microphone in its handset isn't working yet, and what you hear is simply a pretend ring, sent back to you from the phone company's switching center. But then the friend answers, and all worries are forgotten: heart rates and breathing almost al-ways briefly speed up at the excited pleasure of making this contact. There's little question it'll be a woman who answers it: when a man and a woman are the same distance from a ringing phone, it's far more likely that the woman will go answer it. (The most detailed studies have been in Britain, where it's three times as likely.)

The skill of dialing that starts off this whole operation is now wide-spread, but once it wasn't. Dwight Eisenhower lifted up a phone one day after he retired from the presidency, in early 1961, and had no idea what to do when an operator didn't answer. He had last used an ordinary phone decades before, when dialing was still rare: in all the time since, official aides had seen that his calls went through. A holdover from that era of rotary-dial phones is the system of area codes Americans still use. The biggest cities have area codes with the lowest digits because low numbers were the quickest to dial. Thus Manhattan's 212, Chicago's 312, and L.A.'s 213.

The baby looks up from its position on the floor where it has been contentedly occupied chewing the bottom of the bedspread and only occasionally ducking its head in self-defense as grapefruit splashed, wondering why his mother is delaying again. When she sees him look to her she smiles, and he races away, crawling in fun toward the parent's clothes closet, door invitingly open, from which emanate the odors of cedar walls and fine cotton shirts and, especially today—after work on Friday being when everything is collected from the cleaners—the tangy, rich vapor cloud from all the freshly dry-cleaned clothes.

It's hard for a baby or its watching mother, or anyone without a large gas chromatography unit strapped on to their eyes, to see the vast number of toxic molecules in that cloud, invisibly billowing toward them. This is consoling, for it's powerful stuff. Dry cleaning isn't actually dry. Almost everything you leave at a dry cleaners has to be immersed in a solvent to get the stains off. Afterward the sopping clothes need to be dried, but even though the staff at dry cleaning stores do manage to get almost all the moisture off, a certain number of chemicals remain. The most important of these is the potent substance called percloroethylene, commonly known as "perc."

Bring your dry-cleaned clothes home now, and the perc is already vigorously outgassing from them. Its vapors create a high-pressure cloud inside your clothes closet, stabilizing when it's reached about 18,000 micrograms of the molecules in each cubic yard of the air there. Keeping the door closed helps trap the stuff inside, even though a little will seep out from underneath the door. But who can keep their closet door closed forever? Who, indeed, can survive on those flutters of quick openings and then quite unconcerned just-as-quick closings which preposterously lucky beings who suffer no dress-code tyranny—that is, male human beings, who can put on the same suit, day after day, for *years*, and still not be considered dowdy—find sufficient? When it's opened wide, the molecules aren't trapped at all. Combined with the constant underdoor hissing escape, the perc molecules reach levels that average 1,200 micrograms per cubic yard in the bedroom's air and 450 micrograms in rooms farther away in the house.

Once it's inhaled our body's enzyme systems turn the cloying perc into other chemicals. But this only makes it worse, for those breakdown products include such dangerous items as vinyl chloride, chloroacetic acid, and, especially phosgene—the old World War I poison gas. Those get dumped in the baby's bloodstream now, to join the smaller amounts it received, along with the mother, when they had been sitting farther away. Some of the breakdown products are breathed out almost as fast you breathe the raw perc in, but a number of them accumulate in the body, especially in fatty tissue or in breast milk. Anything up to 50 or even 100 ppm in our blood isn't too bad, as we can urinate that away a few hours after breathing it in. It's the amount above that—and how long have you slept with it seeping from an open closet door?—which build up and get lodged inside our body to stay for days. You can measure the buildup in dry-cleaning workers: they don't have much on Monday morning, but by the afternoon the levels are rising and on Friday they are at a peak, taking all weekend to clear away. (The result can be more miscarriages and deformed sperm among dry-cleaning workers, at least in Scandinavia where this has been studied.) Even going shopping in a grocery near a dry cleaner's can bring you some, for the perc sinks into fatty foods such as butter or margarine—studies have found it drifting into packages of food, just from dry cleaning being carried home in a car.

Leaving your dry-cleaned clothes outside to air might seem helpful, but isn't very effective. After six hours of airing the perc level only goes down 20 percent. Emigration is a further possibility, for in Germany and several American states perc is being phased out. Until then though, you can check with the regulators to see that your dry cleaner is one who finishes the drying of your clothes properly: there can be a 50 percent or greater difference in how many of the perc molecules are left on a fabric.

The phone call's done, and the mother lifts her baby from the closet entrance now, her breath blowing extra perc molecules toward his tender little nose and eyes as she bends. She settles back at her chair, the baby on her lap again, and rummages through the cables to set up her laser printer. The baby watches fascinated, and makes fumbling movements, eager to join in. But will she let him help explore?

Different cultures handle this in different ways. In one survey, American parents said what they most want for their kids is for them to be independent and self-reliant. This is fair enough, as it matched what American adults wanted for themselves. In Indonesia and the Philippines however, barely one-fifth of parents went along with that; over 80 percent said quiet obedience was the most important trait to instill in kids. There's a similar difference in Japan, as studies by patient graduate students observing Japanese and American parents at home have shown. When a baby is edging forward eagerly on the parent's lap, the Japanese mothers try to settle it back with soothing or lulling sounds. The American mothers are more likely to start actively chatting with the baby, and encouraging it to explore. Girl babies get it worst of all. The Japanese mothers don't chat to them as much as they do to male babies; the watching students were startled to find that mothers didn't feed or touch female babies or even change their diapers as much. (Lest the American reader feel entirely superior, an experiment was conducted which kept a close look on maternity wards. When both parents were in the room, American mothers were just as attentive to their newborn boys as to the girls. But let the father leave, and then—despite later protestations of total equality!—the girl babies invariably got more cooing, hugging, and general contact.)

In this exploration-encouraging family, the baby's hand is led forward to switch the printer on, then the desktop copier is turned on, too. Tiny clouds of selenium and cadmium dioxide spray over the baby as the motors heat up; then some nitrogen oxide and a little carbon monoxide from the copier's toner; then finally, as the high-voltage electric discharges inside the main motors really get going, a full exhaust blast of corrosive ozone molecules overpowers them all. This is why people who have to use an office copier a lot in an enclosed room end up feeling bad. The baby has had enough, and however much its mother can't understand—for with her adaptations the ozone is no problem, at least now at at these low doses—wriggles desperately to get down. It pads away quickly on little palms and pajamaed knees, squeezes through the half-open door to the upstairs hall, and is gone.

Downstairs meanwhile, a floor below all these activities, the dad is coming in from outside; his lawn mowing finished, or nearly enough. He knows that certain neighbors will have problems with the jungle-like patches left at the far end, but their concerns are easily put aside, for there's a sports program on television, and priorities have to be understood. He strides toward the living room, holding a freshly cut rose in his hand. Desperate chemical messages are still screaming out from the dying plant, leaking behind him in the air, but there are no circling wasp predators to call here; the revenge calls just stream free in utter uselessness, adding though, a little, to the general fresh plant smell.

The living room door is left open as he hunts for a vase. There's some dust on the shelf where the vase is located, and part of him, the sensitive house-caring part, knows he should go ahead and clean it, but another part of him, the male part, realizes that's nuts.

He stands back to get a proper look. It's always been a puzzle how even high surfaces get so dusty, but even at this moment he's adding to it. Unless you live near a dirty power station or big road, most of your household dust will simply be chunks of your family's own skin falling off. Each of us is layered with about five pounds of the stuff—more for the larger members, less of course for the baby—and within a few weeks almost all of the top layers are pushed up and off, as fresh layers sprout up from below. (Most of the growth upward takes place at night, as we sleep.) From just five people growing the stuff and steadily popping it loose, there will be many pounds of skin launched each year. Divided into several billion separate cells, the dried skin fragments flutter everywhere the air currents blow them in the house, arriving downstairs in just half a day of ordinary floating. Any ledges that stick out from a wall are especially good at collecting a terrific biological history, because the stately in-house air seas carrying our skin don't just circulate at random. In this living room, as in most rooms, air rises up in the middle of the room, then spatters sideways along the warm ceiling till it reaches the cooler walls. There, finally, it gets led slowly down, bringing its tiny skin cargo, while anything sticking out—shelves will do, but window ledges are also excellent—is buried under the full impact.

The father sits on the couch in front of the TV. He squeezes an infrared-squirting object—the remote control—and the TV immediately comes on, which is no surprise to the electricians who made it, as a modern TV's circuitry doesn't need the long warming up of earlier models. When the remote control's signal arrives, a switch is released and the electromagnetic signals soon start bombing from the screen into his retinas.

And this is bliss.

A metabolic state unique to television watching now begins: the father's eyes remain busy, constantly scanning, but the rest of his body is slowing, pulse and temperature and muscle tension all decreasing in the average viewer to a level that steadies at 13 percent down from even just ordinary sitting. Any truly intent TV watcher will soon be in the extraordinary position—as physiological measures will confirm—of burning fewer calories than if he were doing nothing at all! This TV body-crash is disheartening for anyone who expected that his body's usual rate of metabolism would use up many of the calories from his last meal, but it's worst for anyone who's overweight to begin with. A check of thirty-one girls watching old reruns of *The Wonder Years,* breathing tubes and other physiological measuring devices attached, found that it was the chubby ones whose slowdown was the greatest.

Are there at least some health benefits from the decrease in heartbeats? A slowing of 10 percent will, after all, save seven heartbeats a minute, which is 420 an hour, or, in three hours of this blissful sports isolation each weekend, a third of a million fewer heartbeats each year. Unfortunately, it's not quite a justification for massive slumping. A man's heart rate is already lower than his wife's, even when he's not lost in TV oblivion. There's a general physiological law that large mammals invariably have lower heart rates than smaller ones. Men's hearts beat more slowly than women's, and women's hearts are slower than kids', whose are slower than dogs, whose are slower than cats, whose are slower than mice. But although this means that the wife pounds out several million more heartbeats each year than her husband, her risk of heart disease is still less than his, because of the general protection from arterial clogging she probably gets from her hormones.

Because the couch-dweller's eyeballs don't join the general slowdown, a condition very similar to a waking dream begins. There's the same inertness of general body muscles; the same exception for the six orbital cavity mus-

Turn on a television or radio and a 740-mph sound wave blasts
out, expanding as a perfect bubble.

cles spinning and tugging the distantly watching eyeballs; the same sudden
spurts in pupil diameter as something of special interest appears in the visual field to be tracked. TV shows are a perfect match: like dreams they bring
us a world where cars start without fueling up, guns fire without reloading,
and you can jump from one scene to another, hair unmussed and not even
breathing hard, without anyone finding it odd. The programs are likely to be
better than expected during four months of the year: February, May, July, and
November. Those are the months that Nielsen Media Research chooses as its
"sweeps" months. Nielsen usually gauges ratings by electronically monitor-

ing a small number of household TVs, but during these months, the company passes out a great number of viewing diaries. The diaries provide a detailed breakdown of viewers. If advertisers don't like the results, the broadcasters suffer. In an attempt to attract more viewers during the all-important months, local news programs are likely to broadcast their most popular features, while the national stations will show their best movies.

High-speed gushes of electromagnetic waves are spurting out from this TV set, as well as from the transmitting station. Some soak right into us, leaving behind about 1.5 millirems of radiation, which is the equivalent of three chest X rays each year—a figure that sometimes can be reduced by sitting slightly off-center from the screen, thereby making oneself less of a direct target. Most of the rippling electromagnetic wave fronts blast through the house walls, traveling immensely faster than the earlier escaping kitchen gases, and are soon hurtling upward to the stars. The result is a great bubble of escaped TV programs now expanding through interstellar space. Its outermost rim carries the grainy image of the first BBC broadcasts from Crystal Palace in north London, while inner bubbles, roaring away a mere 9,200 trillion miles from earth, spread the visages of Lucille Ball, Walter Cronkite, and Ozzie and Harriet.

Lost in the on-screen world, oblivious to the house and lawn beyond, the dad has no chance of noticing the discreet flying monster that is now hovering in the room. Female mosquitoes only come into our rooms to look around a little and fly next to our ears with a horrible irritating whine and then back off a little so we can't find the damned things, because they're tender, caring creatures who are willing to forego any danger to get high-quality iron-thickened water—our blood!—after their eggs have been fertilized. Male mosquitoes, happy-go-lucky wastrels, utterly without a sense of responsibility, never bother us this way. An occasional slurp of cheap, low-grade plant juice is all they need to keep going. Only female mosquitoes bite, because only females need the iron and amino acids and other nutrients that our blood vessels, unfortunately accessible to stab wounds from the surface, come so plentifully supplied with.

The mosquito at the doorway looks clumsy at first, but its body is well designed for tidily harvesting our blood. They're not even being clumsy when they wobble in place. Mosquitoes are so small that they don't have to actually bother with flying in the way we understand it. The air is thick enough at their microsize that it becomes as buoyant as a pond of water would be for us. All mosquitoes have to do is row their way across the air. What they're doing now is simply bobbing in position, tiny wings easily sculling, as they get their attack plan underway. Any male mosquito nearby would hear that sculling, as the whole air pond from our room stretching outward moves slightly in time with the wings. This is what allowed them to find the females for mating earlier. (Female mosquitoes that are too young for mating beat their wings more slowly, sending out slower ripples that the males ignore.)

The mosquito finds the dad by locating the carbon dioxide cloud he's emitting—like all nice blood-filled mammals, humans are crude combustion machines, constantly generating carbon dioxide as waste, which the mosquito can then aim toward. (Tormented caribou herds will hurry into forests, even when there may be wolves there, just to break up the carbon dioxide plumes that could otherwise signal dive-bombing mosquito flocks.) Even a slumped TV watcher will stir now, flapping one hand up, when he hears that distinctive mosquito whine approaching. Mosquitoes will often pause here, hanging in place a little, pumping tube only partly extended, to see what happens next. A second powerful air swish by the human though, more accurate this time, and now only flyers keen for personal investigation of the Mosquitan Afterlife will risk floating so close any longer: it flies away, to hunt for the blood it needs elsewhere.

Deeper into the couch the electronically bonding father sinks, and in his fascination, the physiological embarrassment reaches fresh depths. The basic metabolic slowdown takes over quickly enough, but if he's lugged his feet up onto the couch too, then his breathing slows even more. Gravity no longer tugs his abdomen downward, and the lungs take up a slight outward bulge, leaving a little more air trapped uselessly inside at each breath. The mouth slips open, so ending the effort of dragging air through the nose and pharynx, and when swallowing slows, the drooling that's characteristic of a daytime nap might begin.

It's about as close to being a member of the British royal family as one can get, and the innermost alveolar chambers of the lungs now begin to collapse. Babies that are very premature always have collapsed lungs, because the surface tension of liquids that are on the lung's inner surfaces tugs them together to keep the smallest chambers closed. After about thirty weeks of fetal life, and trustfully forever after birth, a detergentlike chemical is squirted out inside us in fresh quantities about every two hours, day and night, which cuts down the surface tension by about 85 percent, keeping those lung chambers open. Lying perfectly still, however, breathing as quietly as a certain root vegetable upon a sofa, means that the sudsy molecules begin to slip out of position, sliding away from the very surfaces they're supposed to be holding open.

Without some help, the engrossed viewer would have a serious—and soon terminal—problem. Watch any absorbed TV viewer closely and before too long you'll see the sudden burst of activity—triggered by lung-deep nerve endings in place largely for this purpose—that resets all the alveoli and suds. It's often accompanied by a distinctive jaw rotation, shoulder lift, and chest expansion, but you can tell when it's coming on even with your eyes closed. It's the distinctively audible, inward air-gushing reflex we know as the sigh.

Advertisers have to get their show-interrupting pleas across to human specimens in such positions of utterly minimal alertness, but this isn't as much of a problem as it might seem. A lot of money is spent on studies of what works best for sports show ads, and brain-dulled viewers are ideal. The people most likely to act on Superbowl ads aren't the triumphantly hysterical ones, watching from the winning side's city. They are not viewers from the losing city either. Rather, they are the least involved of all: watchers from cities that have no team in the game, gazes fixed and jaws dropped, who let the messages seep through best.

Back in the sunny hallway upstairs, the baby has been sitting upright, enjoying the fresh air after the assaults of his mother's desktop computer and printer, taking stock and deciding what to do. Far ahead is the distance-blurred door of the bathroom, and that'll do fine as a target.

A carpet is easier for tiny padding hands to grip than the slippery tiles downstairs, but it's also—even if regularly cleaned—a stacked museum of fragments from virtually everything the family's discarded or tracked into the home over the past weeks or months. The largest dust items aren't a problem as this pajama-clad explorer heads out, for they land on the very top of the carpet, and are easily carted away with vacuuming. But others can be so lightweight as to only settle in the ultrastill air which fills this house at night. This means that they're on the top of the carpet fibers just in time for hurrying morning feet to really jam them in so deeply that ordinary vacuuming can't get them. But what—along with the bubbling breakfast chemicals and our ubiquitous skin flakes—are they?

If you have a cat, it's likely that you won't think of it as being especially dirty. Indeed cat lovers are known for thinking of their beasts as clean, in evidence of which they point to all the cats careful grooming, and especially the way cats drag their tongues over their fur. But although many cats truly are far wiser than dogs and would never demean themselves with the embarrassing subservience dogs delight in, there is a type of cat (it must be admitted), which despite its near-mystical origins among the ancient Egyptians, has an IQ approximating that of a noodle and is incapable of stopping this tongue-dragging, even when it's already clean. The consequence is that dried cat saliva in the two micron size range is released into your house in extraordinary amounts. A quarter teaspoon of dribbled cat saliva—which one cat is quite capable of extruding in a single afternoon of this lobotomized auto-cleansing—can contain several billion such fragmentary segments. The dried saliva will stay up in the air for hours, and even in a house that has been free from cats for years, it will still be found coating clothes, chairs, doors, bookshelves, computer screens, windows, and above all, having rained stickily from the sky, it will still be coating every square inch of the carpets. If you're lucky, the surface molecules which survive on this majestic saliva rain won't create later allergies in the deep-inhaling baby.

Pollen fragments will also be wedged deep in the waiting carpet vista, independent of the current state of the pollen season outside. Along with any fresh fragments from this morning's assault, many will have seeped in late last night, after a journey that saw them rise from their grass or other release points in the early morning, to be lofted high in the atmosphere by heat currents at midday, only to sink back earthward at 11 P.M. or later—which is why if you have hay fever there's a good chance you'll suddenly wake up sneezing then. Family members are responsible for some of this pollen, particularly

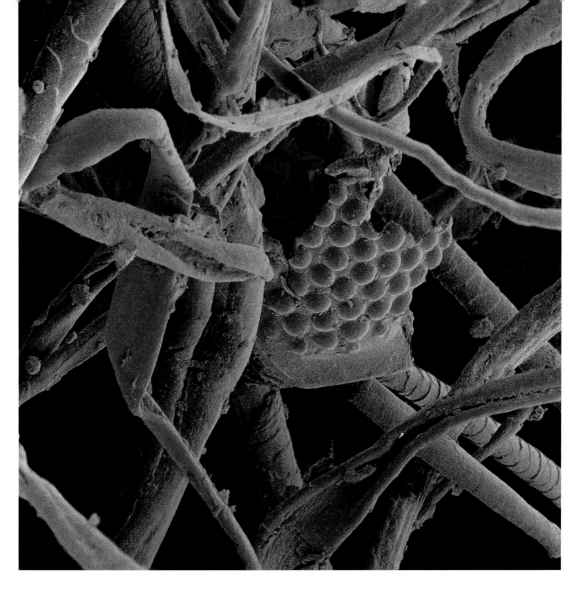

Household dust: nutritiously rich in hairs, skin flakes, clothing fibers, and even—at the center—the intact fragment of an insect's compound eye.

any members that go out a lot, especially in the pollen-rich evenings. It's worst if they have long hair, for that acts as a superb collection net—it's a gauzy filter, just sweeping through great volumes of loaded air—to bring in these treats. (Though, in partial justice, if that family member has hay fever herself, her middle of the night sneezing is made worse when she inhales the pollen from her own draping hair.)

Besides the pollen, cigarette particles, plummeted demodex, food fragments, bacteria, skin flakes, and wedged in deep, all the chunks of shoe-bottom dust that have also been brought in from outside—all this is in the carpet.

When the baby races forward, arms and legs thwacking down, the dust all comes catapulting up, forming a floating haze. Like a great baleen whale, the looming baby monster suctions it all in. The cavernous mouth is filled, and at this point there are two routes of further disposal. Some of it is simply gulped down, sending desperate bacteria, nutrient-rich yeasts, and months-old cat saliva—with only that eight-second delay for all swallows—straight to the huge stomach chamber, with its fatal acid streams. But much of the floating microcargo gets whirlpooled the other way, disappearing down into the lungs.

This could be a problem, for babies breathe through their mouths a lot, which means the dust isn't going to get picked up by the sticky linings inside their noses. They also don't have well-developed tracheal linings further along. What keeps babies from being stuffed with carpet dust and microbes is a population of house-cleaning cells, carted around in vast numbers deep inside their lungs. Grime and rubble come plonking down, soft-landing in the pink inner recesses, and within moments the nearest macrophages start squirming over. These are creatures with all the loving-kindness of Lady Macbeth on a rotten day. They live in the darkened, innermost recesses of our lung tissue, and when the merrily descending bacterium lands, swooping down on what should have been a deserted ideal feeding ground, it's show-time. One bubbling protoplasmic macrophage arm blindly twists forward, then another. A little space opens around the trapped bacterium for a moment, leaving it floating like an astronaut caught in a dimly visible pressure chamber. Then the macrophage starts to digest it alive.

The whole family shares the genetic instructions for creating these armies of marauding defenders, which is why we can inhale so much dust-laden and bacteria-dense air and survive. Sometimes the macrophage unloads fragments of the acid-ripped bacterium right there in the inner lung cavern; more often the macrophage drags itself forward, out of its cavern, till it reaches the wider openings, the very bottom of the bronchioles, where the sticky lining from so far above extends. It pulls itself on, and simply rides the slow escalator up, an apparently inert microlaundry bag. Usually there's a dead bacterium carried inside, but sometimes a desperate, though weak, tugging can be detected in the microscope: the bacterium hasn't been entirely killed, and is still trying frantically to get out. It won't succeed though. Depending on where the macrophage got on, it has days or weeks before it reaches the top. There's plenty of time for a bit more—punishment?—before the ride is over.

Only with stray asbestos fibers that we suction in does the macrophage meet its match. It grabs, it envelops, it sprays, and tries to gulp. But it has no luck, for asbestos is no soft-membraned bacterium, no easy pushover for inner-lung defenders. Asbestos is a rock, and its needlelike fibers, able to withstand the heat of blast furnaces, are certainly not going to succumb to any pathetic chemical attacks here. The macrophage chokes and dies, there in the hidden pink lung terrain, and the asbestos settles in for a long stay.

Any lead that's inhaled or swallowed by the looming baby is equally hard to clear. Lead enters the blood, and although much settles harmlessly inside the bones, some crosses the body's defensive barriers protecting the brain, pushes in to the growing cells there, and kills them. In large doses, as in a house with old, lead-soldered water pipes or lead paint, this can permanently lower IQ, but even low doses are worrisome. Babies are especially vulnerable, because their brains are still growing. The carpet tramped lead probably got there by the baby's own parents or siblings bringing it there. Metallic lead is too heavy to float more than about thirty feet from a car's exhaust, so anyone walking near a busy road or just in a parking lot will get some on his shoes. Carried home it's jammed into the carpet, for any crawling baby to send bouncing loose again. Even the amounts sequestered away in the bones might not stay harmlessly there forever. If the baby ever gets a strong fever in later life, its body will inadvertently let some out of those bone storehouses and loose into the brain-heading bloodstream. (Girls who grew up near lead-spewing factories in Eastern Europe released the old lead from their bones decades later, even in the safe haven of Australia, when the stress of flu or pregnancy pulled reserves from their bones.) Three thousand years ago there was virtually no lead in human bodies, and even now there's only a little bit in Himalayan farmers, so one way to safeguard your kids is to quit your job, grab a guru, and take your family off to the Himalayas. For the less adventurous, you can just take your shoes off when you come in. It cuts the lead levels at home considerably.

The baby is almost halfway through its journey now, bacteria and most dust defended against, when suddenly something else, something quite different and not at all a typical carpet-borne creature, flutters close. It's the mosquito again, flying up from downstairs; utterly lost in this labyrinthal house, and increasingly desperate to get some iron-soaked water from the very next carbon dioxide generator it finds. Its miniature wings crank faster to lift it up above the landing, for at least a quick view of the terrain up here and to see if there's a clear way outside. Its fertilized eggs are in imminent danger of starvation, there in the tiny pouchlike sacs where they're waiting in its abdomen, unless it can find some blood. At first there's only a view of the carpet edge, but there's some sort of puffing air generation device farther on—which could be the long-sought door to outside?—and the energy-fading mosquito cranks one final level faster, slowly lifting higher in this upstairs air.

And there it sights its food.

The mosquito is suddenly tired no longer. It roars up even higher, to properly survey this wholly unexpected bounty. What fair fate has granted it this waiting baby, nice and plump and succulently available for pump drilling, now on the floor just a few feet below? The mosquito wobbles slightly in the air again, determined to be absolutely sure after its problems with the father, but each time the baby burbles in puzzlement a great draft of carbon dioxide gushes up, and with that perfect course guidance around and no repellent to block its detection, the mosquito—flying Dracula-like, its wings wide out—is ready to swoop down.

A ten-month-old baby is unlikely to protest at all at this point, for how can it know what to expect? Instead it's likely just to watch in delighted amazement; each further burble sending out accurate vector location signals, as this fun, vivid toy circles around, then alights on its pudgy belly. The mosquito glances upward once, but the great watching face is happy as Buddha in the distance above, and there's no sign of danger—there's barely any sign of awareness, from this preposterous protohuman—so the mosquito simply gets to work.

It doesn't hurt, not at first, because two very different nervous systems are at work here. The human baby's pressure receptors, as with those on its parents and brother and sister, are widely spaced. If they were closer together, the human brain could be overloaded with too many signals. The mosquito is so small however, that its first slice can be carefully aimed to slip

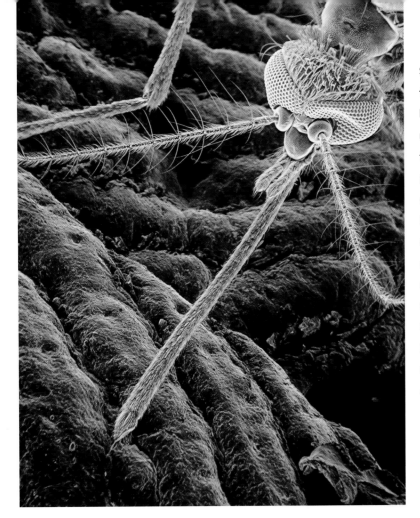

Stretching across the terrain of a human hand, a house mosquito leans in to feed. The tip of the proboscis is sharp enough to slide between our nerve endings, whence the common experience of being attacked without noticing. Only the females bite, needing blood for their fertilized eggs.

between any two pressure receptors on our skin. There might be a slight sensation of tickling, if the mosquito's being clumsy and it tugs on tiny skin hairs as it prepares, but even this is rare.

Once the first slit is made, the mosquito brings down other scalpels, and starts grinding them all into the victim's body in quick rotating motions. To keep the excavation from rebounding closed it starts regurgitating some anticoagulants into the drilling hole. That would be enough if the mosquito had tapped right into an artery, but even when attacking a simple baby that's not a bright idea. The spurted pressure would be too hard to control. Instead, the mosquito has used its rotary scalpels to ease carefully between blood vessels. Only then will it inject the second chemical which it has carried, waiting for this blessed moment, for the entire duration of its flight inside this home. The chemical turns human flesh into a localized mush—it's something like a powerful detergent—and as soon as that's started bubbling away, the mosquito is finally ready to drink up. The main drinking tube had been started into the top of the hole even as the scalpels were still slicing away at the bottom. Now it's pushed all the way down, output nozzles are hooked up, and pump muscles switched on, quickly pulling up all the dissolved human bits. The procedure is effective and with an impressive pedigree: an expedition from the American Museum of Natural History found 90-million-year-old mosquito fossils with mouthparts strong enough to pierce into dinosaurs.

In theory we shouldn't mind any of this, for even a little baby might have a full pint of blood, and the most assiduously drinking mosquito, pumps powered on high, is only going to take a smidgen, a pinch: the merest drop of our total. And although mosquitoes outside of temperate lands have been estimated as the largest animal killer of humans known—responsible, through malaria, for perhaps a quarter of all infectious deaths since the Stone Age—this one is not likely to be carrying any such dangers. But now, finally, self-preservation takes command. The baby's reflexes might be slow, its pressure receptors wide, but it can detect the itching attack of those injected detergents, and it has enough dignity to keep its own body intact. A sudden pudgy *splat* will end the baby's troubles, the borehole that the now-deceased mosquito left behind jiggles closed, though the anticoagulants and detergents left inside will diffuse a little deeper, to itch for hours to come.

Outside the closed bathroom now, and strange sounds of off-key singing, mixed with intriguing vapors of heated bubble bath, are emanating from under the door. The other family members know that it's more than their lives are worth to disturb the fourteen-year-old sister in her midmorning purification ritual. But the baby tries to haul himself up, to reach that enticingly looming doorknob far above.

Inside this mysteriously guarded chamber, powerful electromagnetic

Opposite: Splash a single drop of water in a bathtub and part of it will rebound up, hovering as the oval in this image, before plummeting down for its second and final landing.

waves from distant radio stations are whipping through the bathtub, lightly electrifying all the salts and metals in the warmed water around the novitiate. They're partly blocked by the mud and crumbled volcanic rock of her face mask—always necessary for The Teenager Taking Her Bath—but a number escape through the glass window across from her, traveling as fast as the TV waves from downstairs, and reach the surface of the moon, skimming low over the craters a bare one-quarter second after whirling past her. A few of the frantic electromagnetic waves stay in this room, scurry into the short antenna strip inside the radio, and from there, after being led through the circuits and intensifying in power they come out as the music—as bonding to her teenage cohorts as the dad's TV sports—she needs to be submerged in.

Why does our exuberant bathtub singing sound so terrible? It's not just a matter of the echoes from steam-moistened tiles, though that's the excuse I use. Few people can pay attention to words and tunes at the same time. Words are generally controlled by the left side of the brain, tunes by the right, and our awareness centers find it hard to deal simultaneously with signals leading up from both areas. This is why lyricists and songwriters usually have to inhabit separate bodies, and why yelps, interpolated hums, and general off-key discordance are the best most mortals can hope for. There's also the problem that most of us are immensely far from having perfect pitch. Skill here was once thought to be something you inherited, but when researchers starting looking into why it cropped up so often in musicians' families, they found that it can often be acquired by practice, at least if you start young enough. The trick seems to be to learn the names of the musical notes at an early age and get used to hearing their sequences. Do that regularly, or have parents who harangue you to do that, and scans will show the result: the parts of the brain that accurately recognize sounds grow extra connections.

This safely solo singer's big toe now reaches with practiced dexterity to rotate the hot-water handle. A teenager taking one bath a day, with forty gallons of water used each time, will destroy the equivalent of several barrels of oil to provide the amount of warmed liquid she requires in one year. The heated water now cascading out was once part of huge, icy meteors that hurtled through the early solar system and slammed into our planet, forming the first oceans in a titanic sequence of impacts concentrated when the earth was just 1 billion years old. The particular splashes of water that enter here have cycled through thousands of sea creatures' bodies, and been returned to the sea from rainstorms of the Jurassic and other eras, before reaching this tub. A little chlorine gas comes out from the faucet with it now, added by the local water-pumping company to kill bacteria. Some gas hovers low over the surface like invisible fog banks and is inhaled, to be broken down in the bather's body sometime tomorrow; much of the rest—chlorine being so volatile—slips out through gaps around the window and quickly tries to rise away from the house. As long as it stays combined with the water the sunlight will destroy it. This is the main reason chlorine is used up in outdoor swimming pools so quickly, especially shallow ones.

A colorful bottle is lifted from the bath's edge and something like a tiny paintbrush removed, while the girl goes back to examining the weird objects emerging from several of her body tips. Fingernails are derived from the flexible armored scales of dinosaurs and other early creatures, and have been decorated for about as long as records are available. Toenails, being the most massively armored, take a year on average to grow their full lengths. Fingernails are less demanding, squirmy little things, and whoosh out of a teen's body at a blistering 100 microns—two hair widths—a day in winter and an even more galloping 125 microns in summer. In four months what was once at the base is now at the tip. Parents are slower, younger brothers are faster, but these are only averages. If someone bites her nails, the nails slide out even faster, at 200 or even 300 microns a day. This is when the bottle is really needed, for the amount of nail substance produced doesn't go up to match. A nail that has been bitten ends up thinner and easily cracked. The effect is enhanced by long, family-escaping tub soaks, for the soaking encourages subterranean fungal arms to twist and grow beneath the surface and that easily cracks what's on top. The molten plastic being painted from the bottle fills up the crevices, as well as easily carrying a bright dye.

Teens love choosing their favorite colors, but these are rarely as original as they think. It's hard for chemists to come up with fundamentally new colors, so fingernail polish dyes—as well as those in hair dyes and lipsticks—are regularly reused. Someone with a good memory might notice that the Cyber-Glow Shocker being so carefully ladled out now looks suspiciously like what was marketed as Executive Bold in the eighties; someone with an even better memory, or consulting old photos, might recognize it as Housewife's Harmony from the fifties.

The faucet is rotated off; the once-mighty meteor water halted by the fourteen-year-old's toe. There's a glance at the backlog of textbooks stacked on the floor, but they are easily passed over. American high school students rank highest in the world on U.N.-sponsored polls asking how confident they are that they're knowledgeable and studying enough. As they also rank near the bottom in how much they actually know, something is amiss. One of the biggest differences is homework time. Studies of immigrants show that when they do well at school it's not because they're living in good neighborhoods or that their parents are well off or speak English well or that there are public role models around: it's simply that they've been piling on the homework. (When the whole family sits around the table and does it together—the older kids helping the younger and the parents nearby even if they can't follow the details—the statistical result is even better.) Girls often do better by high school because they worry and tend to blame themselves when they get bad grades; boys, more likely to blame the teacher or the subject or anything but themselves, are less likely to add on the studying that would help. Admittedly the effort needed for getting into a good university is new. In 1920, the majority of Ivy League universities had almost *never* rejected an applicant, with Harvard continuing that tradition largely into the mid-1930s. Rich children were accepted almost automatically; everyone else—aside from the occasional rare prodigy—knew not to try.

It should be time to get out, as the last digit is now painted, skin-tip wrinkles are assuming lunar proportions, and the next stage of the anointed one's preparations—the final skin lotions and dressing needed before re-entering the outside world—are going to take time. Fingertips and toes are the spots that get especially wrinkly because they have the thickest skin. The swollen skin can't spread sideways, and instead is forced upward. Thinner skin, on the arms or legs, is too taut to absorb much water and so doesn't pucker up. It's good to let the polish dry a little, and anyway, it's always nice to pour in a little more bubble mixture from the squeeze bottle on the bath's edge, the one with the pictures of tender gamboling lambs or their like, and then swoosh it around and watch the bubbles rise. The surface that's produced is one of the thinnest substances visible to the naked eye, for a bubble-bath bubble can remain intact at under a single micron thickness, which is far less than the width of a human hair and even smaller than the baby demodex creatures, now desperately clinging to the bath-steamed openings on the eyelashes of this splashing human.

To keep those precarious bubbles propped up, something strong is needed, a long strutlike molecule, and here is where those bucolic scenes of little lambs or calves take on a little more relevance than is generally appreciated. The substance used is somewhat coyly termed "hydrolyzed animal collagen." It's obtained by herding together such sweet and tender little creatures, then smashing their heads in, skinning their bodies, and using electric saws and hydraulic pliers to yank out their tendons. Mashed, boiled, and then skimmed, the result is a perfect, pliable bubble strut. It's a little less expensive to collect such residues from older animals—unsalable haggard horses or lambs that are so diseased their meat can't be used are especially common—so it's modified fragments of their skimmed and boiled body parts that will often be floating around the tenderly musing girl instead.

The soap on her leg is less spectacularly constructed. Raw soap is heavier than water, but air is lighter. Mix enough air into your soap—as happened, apparently inadvertently, to one now-immortalized batch in 1876—and it will float. If that initial botched batch had been only a few dozen bars it would have been thrown out, but as several thousand of the faulty bars had already been made, they were sold as a reduced-price lot. Customers loved the soap, as did the manufacturer, for wherever zero-cost air was filling up the spaces inside, less soap had to be added. The aerated soaps did have a problem at first, in that they wouldn't lather as much as ordinary ones, because all the extra air cut down the amount of potentially bubbling soap available. The solution for several manufacturers is to add hogs' fat. In its raw state this is a less-than-attractive gray sludge, with an odor that mating pigs find delectable, but which most humans would prefer not to bring into intimate contact with their own bodies. Perfumes and odorants are mixed in to mask the stink though, and some titanium dioxide white paint—the same as in the Danish pastry and coffee creamer—is added to disguise the underlying gray color. Rub such a soap briskly over your hands, and a thick luxuriant foam, white and fresh-smelling now, as if by magic, appears.

One more expensively purchased substance likely to be in this tub should be less disturbing even to the greatest sticklers of purity. A teen's lotions and bubble baths are frequently advertised as being pH balanced. What this means is that they have the same number of free-hydrogen ions as ordinary water. This is an easy chemical result to achieve, for in many cases the manufacturers simply add water, at a notably higher price, it is true, than what's pouring out from the faucet.

A final swish of the hand in the water, and the girl gets out of the tub, to finally engage in the task all teen girls must determinedly face at some point in the day, which is to stand on the scale. The whirling dial briefly holds suspense, till finally it stops and she sees, once again, that she weighs too much.

It's a hard fate to avoid, even though this is an excellent time of day to weigh yourself—our body weight goes down to its lowest point every day around noon, then floats up in the evening, swollen with extra metabolized water. Hormonal controls mean that a girl at puberty is going to be pumped up to perhaps 25 percent fat, compared to the 15 percent that an average boy that age will have. The difference is excellent preparation for a future pregnancy, but society doesn't let her put that weight on and be content. American men are now two inches taller than they were, on average, in 1960, and they weigh twenty-seven pounds more. American women are also two inches taller. But how much weight have they put on, to match that energy-demanding height? Just one single, achingly controlled, pound. Only if she could go back in time to Rubens' era, when great thigh-bursting balloonlike shapes were the ideal for female beauty, could she safely be allowed to eat at will, and have the pleasure of seeing any of today's supermodels similarly transported back in time pitied for their scrawny inadequacy. In fact, the modern ideal of slim hips and straight legs isn't especially healthy for women. A comprehensive study at St. Thomas's Hospital in London found that health depends a great deal on where the body's fat is located. If it's found around the waist and stomach it is dangerous, for fat there is regularly broken down and circulated through the blood, where it can clog blood vessels or lead to other problems. Fat that is distributed around the thighs and hips is healthier, because this fat is rarely broken up. (A reasonable explanation for why men have more heart disease, as they are more likely than women to carry extra weight around their stomachs.)

Bathroom scales to keep track of us are relatively new, with the silvery

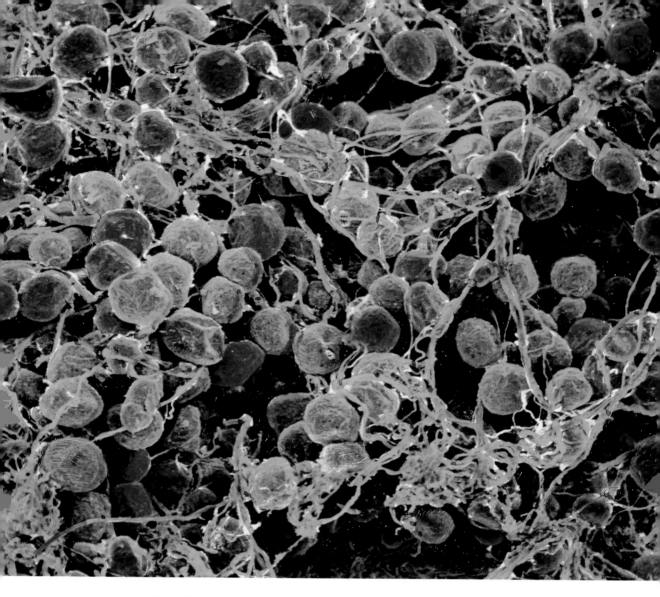

Fat cells. Each cell is a miniature balloon, filling up with liquid supplies.
Thin people keep them partially empty and fat people have the cells burstingly
full, but the number acquired in childhood is with you for life.

Detecto scale of 1927 apparently one of the first. Before that most people had
no accurate idea what they weighed, certainly not from day to day, and so
couldn't torment themselves as much. The two sexes of course torment themselves differently. When women are dissatisfied with their body they hone the
critique down to detailed parts: researchers are inundated with lists of griev-

ously imperfect waists, buttocks, feet, cheekbones, and the like that aren't right. Men are less likely to view their bodies as an assemblage of separate parts with each having to be minutely examined, and tend to worry about their overall strength or reflexes. (The only exception is that men, or at least male undergraduates, are frequently unhappy with their ears—the sole specific object about which their coed counterparts, for once, have no qualms.)

Down the drain from this girl's face go bits of clay, boiled bones, vinyl polymers, trees, alcohol, and yet more paint. All are additional constituents of a good face mask, and the result, when it's finally off, is a wonderful sense of rejuvenation, which lasts for at least four seconds till the girl looks up, to really examine her face in the mirror. She'll put on some lotions later, to get her arms and face looking really good, but for now it's going to be the unadulterated truth.

She can't see the landed microbial flyers touching down from the air, or the deluge-hiding demodex in their sweat-gland caves. But what she does see upsets her immensely: a pimple. She looks around the bathroom, poking through the glass shelf with the aspirin and breath freshener and tanning lotion to find the scrubbing brush and acne lotion— with the extra-powerful formula—she'd left out, carefully in place for such emergencies, before her brother started moving everything. A century ago there would likely have been a certain white powdered concentrate from the Bayer company mixed in there: it was a newfangled drug for family use the company was selling, called heroin. Unfortunately, it had certain problems—repeat sales were *too* good—so market withdrawal came a few years later. It was a different white powder Bayer introduced in the 1890s—aspirin—which lasted a little longer. This longevity was helped by good timing, since later FDA regulations would have kept it off the market, as too many people have stomach bleeding when they use it. (More powerful drugs such as penicillin would have been entirely impossible to introduce. In the 1940s guinea pigs were still being used for testing, and penicillin kills guinea pigs. Only the pressure of battlefield casualties in World War II allowed the drug to be rushed into use.)

The girl finds the small scrub brush and acne lotion. This time she's really going to scrape hard, getting all the infection off no matter how much it hurts, and then she'll put the lotion on, big stinging soaks of it, enough to crush, suffocate, obliterate whatever malicious bacteria—which she picked up at school? from eating chocolate?—have been constructing their career-threatening construction sites on her.

It's a common resolution, but almost always fails, for bacteria are the *least* important source of acne. The triggering cause is the sludgy butterlike sebum liquid that adolescents start pouring out—an approximate half bucket of the stuff each year—from the time they reach puberty. Sebum has no positive use at all. It doesn't guard against sunburn or kill bacteria or stop water loss. The only thing it can do is start acne. It contains fats of a sort that arriving bacteria can break down, and it's those broken fats, not the bacteria, which do the damage. The broken fats are turned into detergentlike blobs, that push out sideways from the face ducts and start digesting what's nearby. It's as unpleasant to view as to read about, whence the agony of the teen trying to hide it.

Why do our bodies do this to us? The latest theory is that it's intended as a marker of puberty: little pulsating signals, conveniently displayed out on the face, saying "I'm Fertile!" Admittedly it's a signal most teenagers could live without—discreet home-page valentines would be a lot easier—but it does give a certified advertising truth: only when you have reached puberty will that sebum pour out.

There is still a role for the bacteria that switch on the detergent blobs but that is something the daughter can't get away from, certainly not here, where although she thinks she's alone and private, isolated from all family intrusions, she's actually more closely connected to her family than ever. An ordinary family bathroom is a wonderful incubation and transfer zone—a near ideal switching center—for the family's diverse bacterial load. The most regularly fed segment of the room's infection core are the thick dangling bathroom towels. Staph and other bacteria wait on the towels, quiveringly alive for many hours. Even if the girl has insisted that no one ever rub their grubby hands on her towel she can still get it, for most families hang their towels on metal rods where the towels are close enough for the bacteria to easily bounce across from one to the other. If the hygienically optimal solution of washing all the towels more frequently is too exhausting to even think about, then simply spacing them so that they don't touch will help considerably. It's the same reason that it's wise not to have toothbrushes touch, though if the ten-year-old boy has the peculiarly ten year old's habit of running his thumb along the bristles of his brush and giving the whole counter top a good spray when he's done, then all precautions short of individual containers for brush storage will be of little avail.

The door handle starts to turn, or at least there's the sound of someone trying to turn it, and the daughter whirls around, furious that anyone could dare to intrude. She calls out, and the baby yelps, then plops back down on its pajamaed bottom. When she bangs on the inside of the bathroom fortress door for emphasis, he yelps again, thoroughly scared, and starts padding at top velocity back down the hall, hurrying desperately, till suddenly, appearing out of nowhere, reaching down for a great swooping lift, is his mother.

She's in wonderful spirits, as she carries her baby back to the master bedroom. Her boss wasn't in when she finally finished her report and phoned him, so she got to leave a message on his machine instead—one of the unsung pleasures of modern life. And how can she possibly do any more work now, when her baby son needs comforting? She triumphantly flicks off her computer and sits with the baby on the bed, holding and comforting and even passing over one of the normally rationed candies he likes, watching as his tears hesitantly stop during the intent concentration needed for unwrapping it. They lounge back together on the big clean bed as she gets out a colorful book to read aloud; they rest in this spotless room, where the windows are always left wide open for air and light and where the pillow in its crisp, freshly changed linen cover is propped comfortingly underneath them.

And where the home's final parallel family thrives.

There's an entire neo-dinosaur landscape of lumbering creatures deep inside the pillows, even in the cleanest of homes. We nourish our pillows with hours of moisture-rich exhaled air each night—a drenching half pint per night is typical—and that, combined with the skin oils and surface skin flakes we can't help but scrape loose, is enough to keep their population at levels immensely greater than the hair-follicle dwelling demodex we saw at breakfast. The demodex existed by the mere hundreds; here, on the pillow, the human family is cozily surrounded by a world—mercifully invisible to the naked eye—with hundreds of thousands of busy inhabitants.

These are *Dermatophagoides pteronyssinus*—the flesh-eating pillow mites. Unlike the cuddly rounded demodex, a microscope reveals the *Dermatophagoides* as hulking armored beasts, with eight legs and massive rhinolike necks. They're also superbly equipped for life inside the pillow— their feet even have flaring pads, like a *Star Wars* desert planet beast, to keep them from suddenly sinking in the soft filling—and despite the forbidding name and appearance, they are actually quite mild.

As it's difficult to see well in the dim light reaching their depths, they signal romantic availability not by crude bellowing calls, but by the polite re- lease of a floating vapor. The targeted one swivels its huge neck to get a di- rectional fix, and then, as gracefully and balletically as an armored monster is able, trundles shyly forward for the hopeful tryst that awaits.

It seems to be a near perfect life, with several generations of these bulky creatures—from gnarled grandparents to thin-walled frisky juveniles—rest- ing, strolling, romancing, or, greatest of pleasures and definitely greatest use of time, tilting their heads up to grab the gently swirling skin flakes tumbling down. But paradise is not for our planet, and there's also one other sort of creature in the pillow: the dreaded, jaw-slobbering *Cheyletus*—a relative giant in this subvisible domain, that lives by tracking down the ordinary peaceable mites in our pillows, and eating them. Let a *Dermatophagoides* adult release a mate-luring pheromone cloud, and this *Cheyletus* will hurry along faster than the intended, to wait, jaws ready, there in the dark, till one of the hopeful suitors lumbers into reach. If the *Cheyletus* can't find suit- ably nutritious adults it'll simply pick off bite-size morsels of baby *Dermatophagoides*.

If this were all that happened, it wouldn't matter much that this odd world is so busily active beneath us. But the mother plumps up the pillow for her and her baby son. Any such plumping, or even any twists and turns we take on the clutched pillow at night, forces windstorm velocity air gusts into that hidden world. The air then whooshes back up, forming great arcing parabolas that rise a full three inches or even more above the pillowcase, loaded with thousands of the discreetly named "anal pellets" each *Cheyletus* has produced. They explode apart in our open air and then float. Since they

Pillow mite. Too small to see, colonies of 40,000 or more pillow mites inhabit the warmth of even the cleanest household pillows. Most are harmless scavengers, living on our scraped-loose skin, though a small number of microcarnivores—hunting these skin-eaters—rampage through their midst under our resting heads each night.

were recently inside the digestive system of an enzyme-secreting arthropod, they're not especially healthy to have floating around, especially as they're exactly the right size to get breathed in.

Adults are fairly well protected by their developed immune systems, but kids, and especially babies, can suffer, as the gut enzymes slip loose from the floating pellets or land on insufficiently blinking eyes. The baby's cells that will later be powerful histamine producers can become oversensitized when enough pellets touch and will stay that way for years. The effect is greater than that of anything else the baby encountered in the hallway: a large epidemiological study in Britain, tracing several hundred Southampton families over the years, found that one of the best predictors of getting asthma as an adolescent was living in a home with large numbers of pillow mites as a baby. It's not just a problem in the bedroom, for the haze of broken pellets floats to all the rooms, settling on baby-exposed carpets everywhere in the house. Regularly allowing the dog or cat to sleep on the bed during the day provides further supplies of warmth and nutrient breath vapor when the animals rest their heads on the pillows, thus incubating even greater numbers of pillow mites. A teenager who spends hours leaning on her pillow during her life-sustaining phone calls—her body thereby acting as a radiant heating coil for the life underneath her—will be guaranteed supremely high numbers in her room.

The results of these studies make people inquire about the price of flamethrowers. The populations are impressive, for the pillow mites have been found in virtually 100 percent of the homes studied, be it in Germany, America, or Britain. There are usually at least 10,000 mites per pillow in the most hygienic of traditional homes. If it is a house where busy professional parents only change the pillow*cases,* but somehow have forgotten to ever rinse, soak, boil, or in the faintest way wash the pillow *itself*—thereby letting the sheltered inhabitants be discreetly fruitful and multiply for weeks, months, or years on end—then the pillows they're using, and considerately providing for the rest of the family, will be home to 400,000 or more creatures. In the distressing estimate of Britain's leading pillow-mite specialist, an unwashed pillow can end up being be stuffed with up to 10 percent living or deceased *Dermatophagoides* by weight over the years. Along with cleaning one's pillows, at least occasionally, it's also good to keep the windows open sometimes. Double-glazed windows and central heating encourage the warm, moist conditions the mites like, and wherever that's kept down, their numbers fall too. Arizona and the Alps work as resorts, partly because their dry air helps kill off the asthma-production machines of such pillows.

The wife hears a call from downstairs, glances at her bedside clock to confirm the late time, and quickly gets up, one soothed baby son happy in her arms. A number of the mites come along too, clinging tightly to wife and baby alike. Habitat destruction is a continual threat. In a pillow not used for five or six months, the entire population of mites will starve. A proportion of the population valiantly travels with us all the time as a safeguard against such disasters. Most die before they're ever brought to another pillow for recolonization. But a number will survive, for the species is photophobic (averse to light) and so without quite knowing why, they do try to hunker down, tiny feet working their way down from any sun or artificial light, to clamp around our clothing fibers—wool and brushed cotton do best—till they're at least partially protected.

We accordingly are the vehicles that transport these city-state populations around from room to room in our homes, as well as carrying them to offices, schools, and one of the most fruitful switching stations: hotel rooms, with their nice, constantly reinvigorated, traveler-awaiting pillows. Here in this house representatives of the several different pillow colonies are being carried on the human family as they begin to assemble downstairs; adults and babies and juveniles and even some of the awful *Cheyletus* predators, though humbled now, and cowering away from the light as much as the others. All the movement is just a vague blur to the diverse *Dermatophagoides* holding on, as the humans collect money and jackets and candy bars and shopping lists; as the baby's bag is rechecked for extra diapers, and the dad tries not to be too impatient, holding the car keys, as the daughter comes down. Even she's been willing to hurry up, for the sake of what's coming next. The traveling mites are going to get a treat today.

This family is going to the mall.

afternoon

4

mall and lunch

Peering out from their car window, in the open parking lot, the family is a unit once again, briefly, at least until the individuals separate. Now they draw on their impressive telepathy of communication systems, including visual signals from over a dozen facial muscles for rough data transfer, as well as those invisible smaller air compressions—words!—for more detailed and precise information transmission.

Our family navigates around other visitors to the mall, careful not to cross into anyone's personal space. Women are usually polite enough, giving a large space to anyone they approach, and don't become too upset if someone comes a little closer. Men demand the largest personal space, especially men from the South, or at least the men at one American university recently, where brave social psychologists bumped a number of different students walking down a narrow corridor. The Northerners showed little effect, but the Southerners almost all had ambulance-foreboding surges of cortisol and other stress hormones in their blood when it was measured right after an encounter. That anger response is also likely to send blood to the hands, where it would provide enhanced sensitivity and strength in handling weapons. (Fear, in contrast, sends more blood to the long muscles of the legs, which is good for quick escapes. Fear also lowers the amount of blood in the skin, thereby reducing bleeding if your legs don't carry you away fast enough and you are attacked. The lack of blood in the skin also explains the famous pallor of fear.)

Luckily, if the dad makes a mistake and does come too close, a commanding, seemingly telekinetic weapon flashes out: a burbling happy smile from the baby strapped to his chest. By deep genetic primings, this usually neutralizes even the most angry stranger in the parking lot today. Face recognition structures seem to be built in to the baby's developing brain, so it locks on to other human faces, and flashes a survival-enhancing smile.

The teenage daughter is strolling ostentatiously behind, making sure that no one, and especially no one under the age of sixteen, thinks that she's somehow related to these people. But as her family strides ahead she soon needs to catch up, or at least not fall too far back. This is difficult in a crowded mall parking lot, for although we're willing to let our bubble of personal space be compressed a little, we really don't like it to be sliced through to zero. The main family groups ahead of her are safe, for if someone walks too closely toward them the whole unit can respond. First there will be a single quick visual flick-scan in the intruder's direction. If that doesn't trigger avoidance maneuvers, there are further defense displays: a flurry of blinks and seemingly quizzical head twists that usually keep potential intruders away. Monitors attached to clumsy walkers receiving such family-multiple attacks show skin sweating and heart rates rising quickly at this point.

For the girl hurrying forward alone it's harder, for now she's very far behind, almost losing sight of the family. She'll have to engage in the one path trajectory that inspires hesitation in even the most brutal of mall-eager speeders. This is to walk not just a little too close to a group of strangers, not even nonchalantly past their edges, brushing a sleeve perhaps, so that their warning blinks and head twists have to be boldly ignored from only inches away, but rather to go right ahead, desperate for speed, and walk right *through* the group of strangers.

This is something we really don't like to do. Social psychologists have filmed poor unfortunates who've had to do this, and then played back the tormented imagery. Approaching a waiting group we almost always first try a submissive posture, of bowed head and gaze firmly held downward. But as we get closer, having to ignore their startlement at seeing us not break off from the collision course, the tension gets too much, and we look up and start doing strange things with our mouths: almost always lips are tightened, mouths are pursed, and excessive head tilts of apology begin. The pulse surges up, reaching 110 on average, and just at the moment of maximum agony, maximum social isolation, right in the middle of the group being disturbed, we usually squeeze our eyes tight closed, in pathetic attempts at total escape.

A car speeds too close to the main family up ahead. A sequence of enlarging car images appear on the family's retinas, and then are cabled inward, initiating a spasm of busy nerve-cell firings in the visual movement centers of their brains. A message is hurried over to decision centers, and only then are the parents' arms finally grabbing for the kids.

Rubber molecules on the bottom of everyone's shoes crater the asphalt on the parking lot surface, as sudden g-forces distort toe bones in the deceleration. The baby's head especially jerks forward, as the father stops fast. This sends the inside of the baby's skull crashing forward, seemingly ready to scrape against his delicate brain. Luckily our brains don't just wobble inside an otherwise empty cranium, perched on the flexible support base of the brain stem. The baby's brain is floating in a slippery warm liquid—cerebrospinal fluid—and when the skull crashes forward, the liquid compresses like an airbag to keep it from impacting. If the dad's stop was exceptionally fast, there might be some bruising to the baby's brain despite this protection, but so long as it's moderate, that, too, won't be a problem. Tiny, long-stretching clearance cells are spread throughout the baby's brain, as they are inside the adults. Normally they wait in standby position, but when there's damage, mobile forms appear—a little like the macrophages we saw in the baby's lungs earlier—and in the hours to come they will wrap around the damaged bits, hauling them away. (There's some evidence that Alzheimer's, in part, can be due to these clearance cells no longer working properly.)

With the feet skidding and brains wobbling, arms need to be propelled outward to keep balance. Miniature angle detectors located in everyone's knees and shoulders measure how much the individual is tipping over, and send signals upward for the brain to do the complex trigonometry needed for correction instructions. Meanwhile, throughout the family, droplets of a long-stored chemical are pumped out from glands conveniently placed over the kidneys, and adrenaline feeds into the renal vein. In a response lasting for minutes until it breaks down muscles strengthen and brain reactions seem quicker. Such an adrenaline response is powerful, but not magically so. The surge makes you stronger, but doesn't do anything about enhancing the bones which actually support you. There have been cases where patients have stepped out of their wheelchairs in the adrenaline-charged excitement of a religious revival, only to have their still-weakened bones horribly shatter a few hours later from having carried the unexpected full body weight.

Sixty years ago suburbs were too sparse to support malls. The parking lot would probably have been a forested or empty lot then. America was a very different country, and if we could suddenly be transported back, the crowd marching around us would seem impossibly different. People didn't wear tracksuits or leisure clothes in public, but had distinct roles, revealed by their clothes. Men wore hats, which also roughly matched their social level. Despite this, incomes were much more uniform than they are now: the current figures, of chief executive officers averaging 120 times what blue collar workers earn, is historically unprecedented. (Even twenty years ago the ratio was only 35 times more.) There were fewer old people sixty years ago; there were also hardly any divorced people, for getting a divorce was strongly frowned upon. There were even fewer left-handed people than today: it was wrong to stand out in any way, and most had been forced as children to use their right hands. There were more immigrants than today, though authorities were fearful that the country would lose its identity because of them; there were fewer black people in most city stores, as official racism meant they were usually kept out. Through it all there was more trust, as polls taken then put the number of Americans who agreed that most people could be trusted at about twice the figure of today. The very trip to a city center where the shopping took place would have been a major expedition, since most people were too poor to own a car or travel far. When U.S. paratroopers made their first training jumps in World War II, they were almost always making their first airplane flights too, as very few people had flown commercially by then.

Slipping behind again, the teenage daughter has stopped to put on her dark sunglasses. Such posturing is understandable: as Diderot put it, "There is only one person in the world who walks, and that is the Sovereign. Everybody else takes up positions." It's also dangerous for her eyes, especially if the sunglasses she's behind are made of a low-price plastic. Photons sprayed out from the hydrogen bomb–making sun make the rest of the family's eye pupils contract. It's a useful reflex to block damage from that ancient light. Visible photons are blocked by the sunglasses, however cheap the lenses, but ultraviolet ones just pour on through. Since they're invisible, the girl has no way of telling that they are there, and her pupil openings won't shrink down for protection. The darker the sunglasses, the worse it is: her pupils will have *widened* in response to the dark cocoon around them, even though the ultraviolet photons are still slamming in. Only more expensive lenses, made of glass or top-grade plastic, will keep the eye-attacking ultraviolet light out. An added blast is coming at her horizontally, reflected from the car windows.

With her sleeves rolled down to heighten the mystery effect, she's receiving less ultraviolet rays on her skin than the rest of her family. In a parking lot miles long this could be wise, to help ward off skin cancer. Here though, with just a few minutes more ambling before reaching the stores, the covering ends up blocking the ultraviolet she needs for her bones. Cells called osteoblasts are crawling inside the entire family's legs, sticking calcium mineral into place. The more you walk or exercise, the faster they work, and the more they stick on. A teenager whose chief exertion of the week was rotating the bath faucet with her toe is not giving those osteoblasts much of a workout. Direct sun rays could help here, producing vitamin D as we saw on the porch. Even fifteen minutes exposure to ultraviolet can produce a lot of the vitamin. In a teenager, it would drift through the bloodstream to the intestine, to ensure that more calcium is absorbed from her food and ends up in her bones. Sleeves rolled down tight stop the process.

The color her clothes are emitting varies a little from culture to culture. In North America and Europe the brighteners added to detergents mean that a blouse we think we're seeing as white actually has a strong bluish tinge. In South America, blue doesn't carry such overtones of maximum cleanliness; instead a reddish tinge suggests cleanliness.

Car fumes spatter against the girl's protective glasses, swirling in an invisible haze around all the grouped walkers today. The heath effects of inhaling these fumes depend to some extent on how much time the family has spent living in a polluted city. Just as with the ozone gusts from the wife's laser printer, their bronchial cells will be working more efficiently than those of families that never had to become used to such air pollution. Many of the inhaled chemicals are blocked even as they start up your nose; what slips through can be further destroyed by backup enzyme systems in the blood, especially the ubiquitous glutathione. Even if a family has moved to an unspoiled suburb, the effect will carry on, to some extent, for months and even years. Any neighbor of theirs wandering in the same parking lot, who hasn't had that regular experience of urban pollution, will pull in the fumes unblocked. Long-time residents of Los Angeles carry lots of the cleansing glutathione around in their blood, even after they've moved; inhabitants of rural Canada, as blood tests show, do not.

Heat from the parking lot surface reflects upward, making the family's surface blood vessels quickly widen, pulling up to a tenth of the body's heated blood near the surface. More than 2 million perspiration holes start pumping evaporative fluids outward. With the air-conditioned car far behind, the family is a marching vapor-spray unit. The newly parked cars add to the distress, for cars are extraordinarily inefficient devices, with 80 percent of the gasoline you just used raising the temperature of the engine and exhaust system, or ultimately ending up as useless friction on the tires or car surface. The family, meanwhile, is eliminating a surprisingly large amount of heat from its heads. A brain weighs only 2.5 percent of its body's total, but the head pumps out 20 percent of the body's heat in adults and an even greater percentage in kids.

Everyone's suffering, though as usual it's the weaker sex that's suffer-

Sweat droplets on a hand. The amount seen here is what's produced after an hour's exercise.

ing more, the one which constantly needs to be reassured and consoled—the men, of course. Women are skilled at sweating: their moisture droplets come out on their skin in small, easy-to-evaporate droplets. Men, on average, pop out larger sweat globs that are more likely to drip off, producing impressive splat patterns on the ground perhaps, but doing little to cool down the mighty predators. The baby's swaddled in its belly-thick fat, but infants and young children can still grow the optimum number of sweat glands they'll need as adults. Japanese who are born and raised in the tropics end up, by one count, with 2.9 million water-releasing sweat glands, but their cousins who spend their childhoods in cooler lands make do with only 2.1 million. There's a lot of potential cooling from 800,000 extra glands. That's why people can be uncomfortable if they grow up in a cool climate but move as adults to a warm one.

The family enters the cool, air-filtered sanctum of the mall. A great deal of thought has gone into designing the entryway. The entry has to seem inviting, so people will be happy to come in, but it can't be too inviting, for then they'll linger and block it. There are likely to be benches attractively set down just inside the entrance, about as welcoming as a shopping arena can be, but look closely and you'll see that they're usually placed so that a sitter will have his back to the main traffic—a position that's psychologically all but impossible to keep for long. Tiny muscles tighten up on the family's forearms and back, forming the miniature air-slowing windbreaks we call goosebumps, as they try to orient themselves in the suddenly chilled air.

Uncomfortably hard tiles clack under the family's shoes for the first yards in, and this almost certainly makes them walk faster—about 10 percent faster, in one study—than if there had been a carpeted walkway. To be extra sure that there's no slowdown, faint clouds of ionone and other molecules are likely to be spraying out from the recessed cooling ducts further along. Ionone is easily produced—it's just the result of breaking down common carotene molecules. It gives off the delectable odor of fresh hay. There are companies that sell these chemicals in liquid vats for mall managers to use in air-cooling systems. The unembarrassed can usually confirm this use by going up to an exposed duct—perching on someone's shoulder is sometimes necessary—and taking a close-up sniff. Farther on there will be different vapors. Shoe stores are likely to spray out a liquefied leather extract to draw in passing browsers; for sports stores, wisps of floral essence work best.

As they walk, the family almost certainly follows a reflex to veer toward the right. Children do it and adults do it—whether they're left- or right-handed—and it's so powerful an instinct that stores often pay higher rents for locations at the right-hand side of the entrance where the speed-controlled temperature-dazed shoppers are most likely to be piling in. Fluorescent lights above add to the initial disorientation, for these lights exhibit a constant slight flickering, which can create detectable nerve-cell spasms inside the brains of our family. The problem can be avoided when the fluorescent lights are perfectly adjusted, but with the hundreds of lights around, and janitors who have other things to do, that's not likely to happen.

To get upstairs, the family heads toward the elevator. The mall managers may keep one elevator on the ground floor always open. People march in, and even though they wait there exactly as long as they would had it not been standing open, they almost always feel better huddling inside rather than out. The next trick is to appease restless waiting shoppers is to put mirrors on the wall of the elevator. We all tend to glance at ourselves when there's a full-length mirror nearby, be it the discreet side-glances at hair partings or shirt tucks which men are allowed, or the full-out, face-to-the-mirror, inch-by-inch makeup examination which certain Scarlett O'Hara wannabees engage in. (The observations are universal, but the admission of them isn't. In one comprehensive British poll, virtually all men said that they never looked at themselves when they passed a store window. But observers discreetly stationed at a number of mirrored posters observed a rather different truth: men were twice as likely as women to glance at themselves as they walked by, though, in mild fairness to that vainer sex, they were briefer in their mirror-assuring glances.)

A final distraction to impatient elevator waiters—though this one is expensive—is to staff an information desk across from the elevators. Up to one-third of the people coming by will then be curious enough or nervous enough

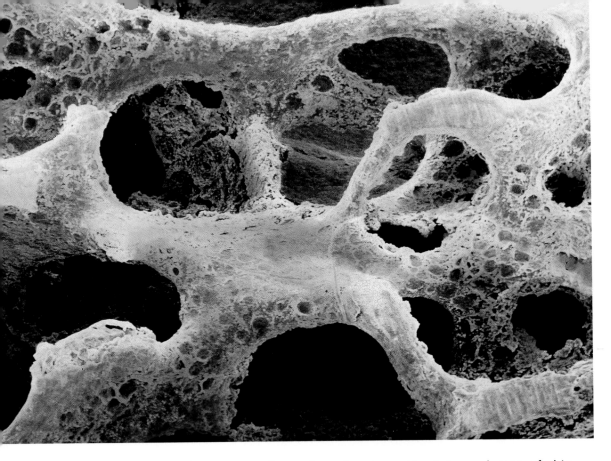

The crumbling pits of osteoporosis weaken the strength-giving columns of this bone, in contrast to the healthy, calcium-dense bone on the cover.

or just polite enough, to stop and check on location details even if they already know perfectly well where they're supposed to be going. That further reduces the buildup of people waiting for the elevators.

(Will we accept anything when we're being led around? At one airport, passengers were annoyed at how long it took to wait for their luggage. The walk to the luggage area took one minute, and the wait there took eight minutes. The solution was elegant, if disturbing when you think that most of the people who fell for it can vote: the airport staff laid out a path that incorporated arbitrary twists and turns so that it now took eight minutes to get to the luggage area. Only one minute had to be spent in front of the empty luggage carousel. Even though the total wait was still the same, there were now fewer complaints.)

The boys are fidgety here in front of the elevator. The friend reaches into his back pocket, and extracts a plastic-wrapped storage device that has been there for several weeks, and contains a mixture of pigs' fat, rubber, Vaseline, Elmer's glue, and other tasty substances. The son looks over with interest, despite the parents' disapproving glances, and as friends are nothing if not generous at age ten, he is offered a piece: it's chewing gum.

Most substances that you can put into your mouth fit into one of two categories. Either they compress and break and become a mushlike slurry you can swallow—apples, steak, and toast are notable in this category—or they don't compress, and you would be strongly advised not to put them in your mouth at all. Those are the nonfood items, such as rocks, twigs, and their like. It's very hard to get items which are in between, both destructible yet not terminally compressible. The engineers of the modern gum product, however, have managed it. Their solution is to get substances that are soft enough for chewing—that's why all the Vaseline and animal fats such as lard and beef tallow are there—pour them into a tough rubber matrix, and manipulate it in such a way that no one is going to guess what's hidden inside. Something like the Elmer's glue we saw on the stamps is used, but there are also dollops of soap, polyethylene (the stuff that makes up plastic bags), and even some soil stabilizer to keep the gum bound tight.

As the boys chomp away, Vaseline and sheep's fat drip out. It is all swallowed, and will be worked on by the liver's detoxification systems in the hours to come. Sugar oozes out too, but this is also defended against, at least in part, for teeth aren't the passive chunks we generally take them for. Their surface chemicals constantly pull in fresh minerals from saliva, and use those minerals to help toughen the regions coming under acid attack from bacteria. At the same time, your saliva does its bit to help. Saliva is not a simple watery fluid, good for spitballs and other boyhood treats. It's constantly changing through the day, as the inside of your mouth monitors what's going in, and the saliva glands readjust their chemical output to help. Here, with the boys doing their best to harm their teeth, the glands detect the need for something to destroy acid. Simple bicarbonate of soda is excellent for this purpose, and so, within a few moments of the gum chewing starting, that's what the saliva glands pump in. It binds with and neutralizes the acid that the bacteria in the mouth are producing.

Saliva glands are truly intricate chemical factories, with a blood flow

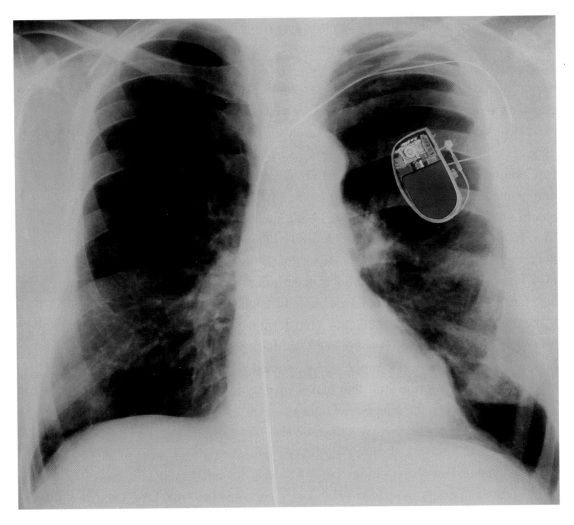

Useful and silly additions. *Left:* an electronic pacemaker, crucial for maintaining the beat of a wavering heart. *Right:* a bag of silicon implanted in a female chest wall, defying the laws of gravity.

that's ten times greater than an equal mass of actively contracting arm or leg muscle would produce. The boys' saliva glands are pumping out another chemical, called sialin. This is even more cunning, for it floats over to the teeth-hugging bacteria already in place and subverts the way they use amino acids to make energy. As a result, those bacteria start producing alkali chemicals, that neatly quench most of the threatening acid.

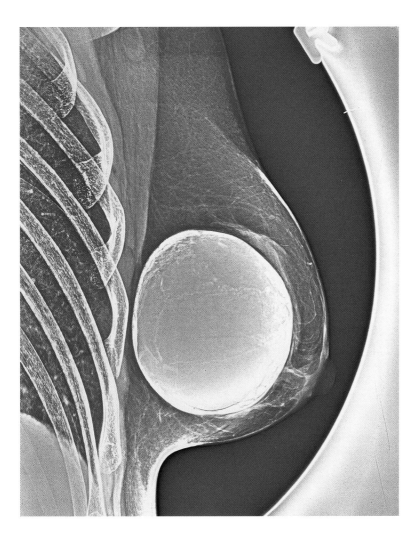

A chime rings, and the family steps forward. Even in enclosed spaces we try to keep our personal distances, and here the girl has one advantage: a single woman is almost always given more space in a crowd than anyone else. (Though as social psychologists have ungenerously observed, the more plain looking she is, the less the deferred space.) The elevator doors close, and the parents and everyone else trust that the little cable above will lift the elevator to its destination. In 1854 the original Mr. Otis stood inside an open elevator three stories up, and had visitors at a New York exhibition cut its cable with an ax. It started to fall, and everyone shrieked, but then the little retracted rollers he had invented sprang out to grip the metal tracks, stopping the elevator, and everyone applauded. A device using the same principles is on all elevators now. Even simple escalators were considered fearful devices once, which is why when Britain's first escalator was installed at Harrods in 1898, a uniformed attendant was assigned to stand at the top with a glass of brandy at hand to revivify any traveler overcome by the mechanical ascent.

The elevator box jolts up, and as it does music plays, almost unnoticeably, which is just what the American Muzak Corporation likes. The sounds they project over an estimated 400 million people worldwide each day are designed *not* to be noticed. High notes are screened out by Muzak engineers, as well as low notes and volume shifts and even most minor chords—anything that might hint at some world beyond the major-chord mush, the amiable hushed bounciness, which is all that is left. The result is a product unique in musical annals: a harmonized sound sequence *entirely without musical content.* It is even transmitted in mono—one of the last nonstereo sources around—to enhance the screened-out simplicity.

Our biorhythms fade a little twice in each day. The major tiredness collapse occurs about four or five P.M., and the other, a lesser dip, at about 11 in the morning. The Muzak that's sent out from the company's North Carolina headquarters is designed to counter these lulls. The sequences that early-morning shoppers will hear are fairly slow, but speed up toward 11 A.M., then slow and later speed up again to counter the afternoon dip, too.

It works, as Muzak corporation brochures cheerfully point out. Cows produce more milk and army staffs watching radar screens make fewer errors and cardiac patients in intensive care units recover better when they hear these counterfatigue sounds at these times. The ten year olds chewing gum, not quite realizing what is happening, will find their rate of chomping goes up slightly.

Teenagers especially dislike it, so when the owners of a 7-Eleven store in Thousand Oaks, a city near Los Angeles, found that they couldn't get rid of teenagers hanging out there, they took the gloves off and used a "Muzak Attack." The store let the warbles and glissandi and woodwind arpeggios pour happily forth. The teenagers vanished.

The family steps out of the elevator, to be met with a blast of microbe-rich air, far denser than anything they experienced at breakfast. It's supplied courtesy of the great number of nonfamily members here: all these wheezingly outpouring fellow shoppers. It's also more diverse than what the single ten-year-old friend could offer, containing a baroque bestiary of diverse, living species. There are some seriously dangerous items. About 10 percent of the country's population permanently carry around the dreaded meningitis bacterium in the backs of their throats. It staggers out as they talk, and when you walk by, there it is, hovering as a dank haze. Isolated but viable microbes for pneumonia and herpes simplex type 1—it sprays out from the liquid in any cold sore blister—and strep throat are also likely to be floating loose in an ordinary mall corridor. (You're entirely safe from the AIDS virus, though, for that one crumbles apart after only a few seconds in open air.)

The miniature windshield wipers of the family's eyelids start sweeping faster than usual, to clear a path through the dense clouds, but some of the bacteria land anyway and then it's time for serious defense. Everyone in our elevator-exiting family, even the baby, now starts pumping concentrated lysozyme chemicals into his or her tear fluid, that bursts apart the bacteria's walls. Everyone could be in more trouble from the bacteria that don't get intercepted at the eyes. Microbes are so small that they evolve in a matter of hours, while humans are so galumphily huge that we take a quarter century or so to reproduce. Without some equally small and fast parallel world operating inside us, we'd be sunk.

Enter the Memory B's to help.

Almost all the white cells we used years ago against any particular infection will have died out, but a few—the Memory B's—survive, forming a wondrous miniaturized museum. For years they travel, little counterpart families, gliding through us. One last remnant of a fever from four years ago bumps silently past the memory cell of the throat infection everyone had seven years ago. The very bodily terrain they knew changes around them: first our skin and then our muscle cells are replaced; in time, years passing, even the original stone-hard bone cells that were around when the memory cells were created will be built afresh. Only the Memory B's survive: their original roles, their very presence, soon forgotten by the humans they're wandering within.

And then, suddenly, they're needed again. The bacteria floating in this corridor have probably been let loose some time before, somewhere in the city or suburb. If anyone in our family had been infected by them, even years ago, then probably the whole family was exposed to them, and so each member will be carrying some Memory B's from that particular event. When the bacteria come in now, the old Memory B rejuvenates, and this time there's no need to wait while the whole paraphernalia of ordinary white blood cells is built up from scratch. Everything gets a head start. By the time this human family leaves the mall—before there's even time for any of them to notice what's happened—that infection is being crushed.

The only thing that makes the job harder is if too many of the people you pass have been taking antibiotics. Those kill bacteria of course, but they never kill all the bacteria in the body of the person who's taking them. A number of bacteria survive, including a few now likely to be strangely mutated: entirely resistant to the drugs their host has been taking. The genetic information that carries this resistance doesn't stay locked inside those few mutated bacteria: instead it concentrates in tough coils that are easily pushed out through the bacteria's surface, to be absorbed inside other, wholly innocent, bacteria nearby. It could be in the host's body; it could be in you. When that happens, passed to you in the public air here, not only do you have a new infection to deal with, one that your Memory B's don't have the faintest clue about, but the antibiotic your doctor might order for it is probably not going to work either. It doesn't help when doctors prescribe antibiotics with reckless abandon and start off the chain. But who, truly, feels comfortable leaving a doctor's office without a prescription for something, however useless, that might help? When whole families do it, the result is that we regularly end up walking beside people loaded with these mutation-inducing antibiotics, every day.

The herd of floating microbes that don't hit us on the first pass in these mall corridors have a good chance of getting us on a second go, and so make our immune system go through the whole rigmarole again. This is because mall managers will do almost anything to keep the air clean as long as it doesn't cost too much money. This means that the easiest solution, of switch-

Vitamin C. The monolithic slabs shown melt in the human body; large doses are avidly consumed for resistance to colds, but end up quickly floating to the bladder for disposal.

ing the ventilation system on high to get rid of all the floating junk by blowing it outside, is unfortunately out. It would cost too much to cool or heat the excess fresh air they'd have to bring in to replace it. Most malls only change about 10 percent of the air each time it passes through the huge fans hidden in the engineering units: the rest is simply shunted back down on us, having been blown uselessly around, increasingly loaded with carbon dioxide and sweat particles and wriggling hair yeasts and serious infections as the day grinds on. Instead there are filters and efforts to keep the lightbulbs wiped clean, and more use of those liquids that can be pumped through the air ducts to give the impression of fresh open air or lawns. Monday mornings tend to be the least polluted, but Saturday afternoons are the worst, as the buildup of exhaled gases bobbing along in the air around you in this too tightly sealed city can constrict blood vessels or change their flow so that headaches result. Drying the air would help, but then people would get static electricity shocks, so it's back to the aerial-soup-preserving humidity for all.

Smokers in the mall make everything worse, even if the mall forbids smoking. A little poisonous carbon monoxide still trickles from their lungs for up to an hour after their last smoke. Since this has a buoyancy almost exactly equal to ordinary air, it bobs at chest-height breathing level even after the smoker has walked on. You'd be better off in an airplane, especially if you can afford good seats, for jets have compressors connected to the pure atmosphere outside that get rid of 50 percent of the used air each time it's pumped through.

The family stops at a particularly intriguing window display; bravely breathing in as they comment. Everyone seems to be looking at the same display, but brain scans would show that they're literally seeing different things. For vision isn't a passive operation letting everything ascend, rawly unfiltered, into our minds. We'd suffer an impossible overload if that happened, which is why we regularly block out what are likely to be unimportant sights. If the teenage girl, for example, is really familiar with the fabrics here, then the visual impulses flick inside her, but her memory cells squelch some of the fabric signals, reaching over *before* they rise to her consciousness. For different family members, different visual sights are likely being suppressed, as scanning devices fitted to macaque monkeys have eerily shown.

Microcubes of dense ground-level air stream out from their middle ears—a leftover from the elevator ride up—but that's no problem. Only if there's a distant approaching thunderstorm is there likely to be the type of changed air pressure, or the distant, inaudible, ultradeep sound tremors—that creates a general, inexplicable unease.

A familiar couple appear in the distance, for one of the partners is a colleague the wife works with at the office, and because she'd rather do anything than face them now on her free day, she has no choice but to lead her family on over, and pretend that she's delighted, overjoyed, at their intrusion. Soon they are standing in odd, rigid positions; strangely garrulous and jocular, their faces twisted in muscle-tugging grimaces. Sinuous forearm extensions and a powerful pumping action—the handshake—are initiated, as the parents burble overenthusiastic greeting sounds. The kids awkwardly stiffen, knowing they're about to be questioned—with wild inaccuracy, and their mumbled corrective replies registered not at all—about grades, and favorite TV shows, and anything else the couple can think of to fill the greeting void. Even the baby waggles its yellow ducky enthusiastically, as it ponders these strange adults around it.

The handshake that formalizes this greeting is something we take for granted, but has several times become controversial. In Mussolini's Italy you could get thrown into jail if anyone saw you do it. Only the stiff-armed fascist salute was officially allowed: everything else was considered an insult to *Il Duce*. In the mid-1800s, English aristocrats were appalled at the American habit of shaking hands. One English officer observing the Confederate Army couldn't contain his distaste at the way General Lee allowed subordinate officers to shake his hand. George Washington, however, rarely shook hands in office, feeling that it lowered the dignity of a president.

Conversation begins with the colleague. When two people talk, one of them will slightly change the basic pitch of his voice to match the other. But which one? Usually it's the lower-ranking or less confident person who does the shifting; the dominant one, however casual or egalitarian he might appear, simply holds that part of his voice constant, as the other person races up or down to match. It's hard to control consciously, for it's not the whole spectrum of the voice that shifts, just this one particular part, in the region below 500 vibrations per second. This is why it reveals who thinks himself on top. When the conversational styles of a number of well-known people were ranked, based on how much they forced television interviewers to match their basic voice frequencies, Barbra Streisand and Bill Clinton came out with a quite dominant 0.80, while Elizabeth Taylor had an even more swaying 0.84. None of the Americans ranked achieved the incomparable assurance of the British politician Paddy Ashdown, which is perhaps understandable, as before entering politics Ashdown was an officer in Britain's SBS, broadly equivalent to America's Delta Force—not an occupation that tolerates submissive hesitancy. (The lowest of all individuals ranked in any country, quiveringly jumping to match whoever he was talking with, was one Daniel Quayle, who barely made it into positive figures at all, with a ranking of 0.09.)

Along with the handshake, the fixed smile has to be kept in place, which presents more of a problem, for there are definite pitfalls, each of which has to be painfully avoided. The first step is to keep the muscles around the eyes pulling hard, to avoid the giveaway of an ostensibly eager grin that fails to reach flat, uninterested eyes. The next step, even more difficult, is the struggle to keep both sides of the face smiling evenly. An ordinary smile is produced by pathways descending from the emotion-generating limbic system in our brain. The pathways are naturally symmetrical and ensure that both sides of the face pull in unison. But an artificial smile has to be generated from the more revealing motor cortex in the brain. This travels via a different network of brain-cell cabling to the centers that trigger the face muscles. The nerves that control the right side of the face frequently don't work with as much power as those controlling the left in this cabling. It's a brief imbalance, usually lasting just one-fourth or so of a second till we can stabilize and get it right. A strange gender difference pops up here. In lab tests the quarter second giveaway is usually too brief for men to notice, but women see through it almost every time. The fetish to insist on smiling at all such occasions is not universally shared. Japanese families often discourage their children from smiling, as it's generally thought to be a sign of lower intelligence and insincerity. Examination of the stiffly held smiles of certain politicians might suggest Japanese parents have a point.

"AAAA-CHOOOOO!" Everyone steps back, as the colleague sneezes, then tries to apologize, and sneezes again. The baby shrieks, utterly unprepared—it was born too late to get into the family's most recent Memory B cycle—as a gush of live nasal viruses pours over him. Cold viruses are especially well designed to travel this way, as they live best at about 3 degrees below body temperature, which is the temperature of air-cooled nasal caverns. The viruses have also evolved to be less than fatal to their human launch platforms, for otherwise we would be collapsed under the covers at home, gurgling pathetic requests for orange juice or a new cable channel, instead of helpfully staggering out of the house to sneeze them on their way to new, resource-rich nostril abodes. (That's why water-borne diseases—where the mobility of the targeted victim doesn't matter—are more often fatal than airborne ones.)

Cold viruses don't only transfer directly through aerial sneeze clouds. Surprisingly often, researchers have found, the viruses are sprayed, indirectly, onto hands, and it's those hands that carry the live infection. When you have a cold you're constantly reinfecting yourself, for although your immune system is killing many of the bacteria in your nose, you are constantly bring back fresh supplies by touching contaminated tissues or trouser pockets, and then dabbing them back into the nose for another round.

The wife isn't going to lose this excuse to escape though; she lifts the baby from the dad, and says she'll just find a restroom to feed him. Only when she goes off without taking the bottle does the colleague's wife realize, aghast, what she's intending to do: this executive woman, in our era of high-tech medicine and sterilized bottles, is actually going to breast-feed her child.

It's a wise choice. A colleague so intemperate as to sneeze all over the place at the mall on Saturday has probably been sneezing all around the office during the week. The mother will have ingested some of those viruses, processed antibodies against them, and will now send them glugging out in the breast milk. It's a marvelous mix: along with the antibodies, there are also growth chemicals the baby needs. In the weeks when the baby is finalizing its visual circuits, a retina-enhancing chemical is added to the breast milk. When the baby passes that stage, the retina chemical stops. The overall effect is impressive: the IQs of eight-year-olds in America who were breast-fed average eight points higher than the IQs of those who were bottle-fed.

Babies have evolved a number of skills to ensure they get this nutritious help. One is rather extreme: before modern hospitals, excess bleeding after labor could easily be fatal. But if a mother reaches for her baby and breast-feeds at that time, a cascade of hormones will be triggered that make her uterus contract, greatly reducing the bleeding. A breast-feeding baby moves its eyes, radarlike, to focus on just the range (about eight to twelve inches) needed to see the mother's face when feeding. If the mother's face slips from view, the baby's pulse immediately goes up, and it simply starts fast systematic side scans till it finds her again. When the mother's found again, the pulse settles back down.

At ten months there's real feeling involved in doing this, but a newborn's cortex (the higher reasoning parts of its brain) is not very developed. The smiles and eye-tracking and wiggling are controlled by its much lower brain stem regions, through instincts prewired before birth.

Into the food court, the boys are first off to select their lunches, but although they stand right beside each other, eyes similarly wide before the burger grill, their bodies react in different ways. The family's son lusts after the meat, as all ten year olds do, but his body isn't going to release insulin till after he's eaten it. That makes sense, because insulin lowers blood glucose, and after a meal there's going to be plenty of extra glucose around. As he looks at the food now, insulin levels unchanged, he doesn't feel any different from how he did a moment before. But fat people are stuck in a horribly different world. The chubby friend is quite probably leaking insulin already, just from his first look at this food, and that means his blood glucose is going down already, fast, and so he's not just hungry; he's desperate to eat.

There's a forced intake of breath as he tries to resist, but nothing about being overweight is easy. Fat children are only safe in school up to the age of five or so—kindergartners will sit as closely to a fat child as a normal-weight one—but by the age of seven, children will avoid sitting near even a cardboard cutout of a fat person. It continues. American teenagers prefer pictures of "people with facial disfigurement, a hand missing, one leg, blind, or paraplegic" to one showing a fat person. A fat person is less likely to do well on college entrance interviews, and less likely to earn a high salary than a thinner person with the same qualifications. Fat women in particular suffer, having household incomes that are $7,000 below the average.

Even when fat people know what's in store for them, it's hard to change. The chubby friend is likely to have more fat cells than the son, waiting like partly empty microballoons around his body. Though he may have been very very good and not eaten enough to load them full in the past weeks, their very

Not E.T., but the common intestinal tapeworm *T. saginata*.
Thriving in undercooked hamburgers, the head openings allow it
to dig in tight in its domain, be it the food it travels in, or the
human intestine it reaches.

number means that he suffers a greater pull of hunger now—that hunger
seems to depend in part on the sheer number of waiting fat cells someone
has. His parents might disavow all blame, saying that they had no control
over what he inherited, but that's not entirely true. Few people are born with
extra fat cells. You can't grow new ones as an adult, but you can, very easily,

as a baby. When the friend was very young, his parents, however well intentioned, were probably loading him with more food than he needed. (Some evidence that it's not genetic is that a baby adopted into an overweight household, not biologically related to the parents, is likely to end up with extra fat cells like everyone else there.) Such a bequest is an especially unkind one, as once you've put fat cells on they don't come off: the number can grow, but will never decrease.

Burgers are selected, and the friend—despite all secret vows that he is not, absolutely not, going to overeat today—somehow ends up asking for two. Napkins and glasses and little folded packets of salt need to be collected, and here another basic difference between families comes in. Which of the two boys goes off to get the items, and which one waits? One famous study at a summer camp photographed a group of adolescent girls playing tennis. Everyone would hit the ball when it came to her, but after that the chubby ones would simply stand stock still and wait. Only the girls of normal weight ran around getting in position for the next shot. The film's frames were counted. Even playing singles, the chubby ones were utterly inert about 77 percent of the time.

The boys sit down and dig in; hands almost quivering as they lift the wondrous grilled meat slabs mouthward. The table that's been selected is unlikely to have been taken at random. Food court managers can usually manipulate where we sit by making one area slightly dimmer than others. People have a reflex (it's called the "moth effect" among planners) to settle in to seats that are in the slightly dimmed area, but which face direct light. If the lighting pattern is changed, the patrons will shift to different benches so they can face what's now the brightest light.

The parents, glancing back from where they're still waiting, aren't sure the boys' burgers are the healthiest of foods, but since they are prominently advertised as being 100 percent lean beef, it can't be too bad. Can it?

To understand how fast-food hamburgers can be sold at their extraordinarily low prices, one has to venture away from this mall, to the most expensive restaurants across town. There diners are likely to be ordering exquisite Chateaubriand or other high-priced cuts. These are produced from the sort of hulking beef cattle that farmers are justifiably proud of: fleshed up on a high-protein diet of plenty of refined grains. But these grain-fed cattle also develop a layer of thick hanging fat below their chests—it's the chunk of soft tissue, hanging like a chest-front jowl that you can see on many prime steers. It would be a terrible waste to throw that huge fat chunk away.

The boys bite in. The meat that fills up most of their burgers did *not* come from any such grain-pumped cattle. That would be too expensive. Instead, it's largely supplied by scrawny, often underweight cattle, that are usually left to forage for whatever scrub they can find. (At one time it was often tropical lands—local peasants forced away—that were used for grazing, though responsible retailers have since disallowed that practice.) Unfortunately the meat from those cattle suffers from the minor problem of being nearly impossible to chew. It's too stringy. Only when it's squeezed together with the otherwise unusable yellow fat taken from the chests of the healthy big cattle does it become soft enough to eat.

The result is that kids loading up on budget-priced fast-food burgers end up subsidizing the wealthier diners. It's not the most equitable system in the world, but the people making the laws are more likely to be in the steak-eating half, and don't especially mind. To make it work you do need to get around the fact that most people wouldn't like knowing their burgers are loaded up with leftover fat wads. But that's easy. The laws have been adjusted so that a hamburger can be labeled as "lean," even when it's notably high in total fat.

Just to make it worse, fast-food burgers usually need to be bulked up with other bits of the originating cattle, because you can't make a truly filling burger with just stringy muscle and excess fat. Skin, pancreas, head muscles, gristle, and diaphragms can be included in unlimited amounts. With certain less attractive fragments—rectums, testicles, lungs, and spinal cords—proprieties are observed. Only "lesser amounts" of these can be included in a burger labeled as 100 percent lean. Whatever goes into it, a burger is immensely wasteful of water. It takes 200 liters of water to produce one pound of potatoes and about twice that to produce one pound of wheat. But it takes 45,000 liters of water to produce a pound of hamburger meat. Almost all is used for growing the feed, with only a small amount needed for the cattle to drink.

A gluey red substance drips down from the bun-squashed concoction, as the boys complete their bite. There are not a great deal of tomatoes in this ketchup—in England, for example, the legal minimum is now down to 6 percent tomato solids—but that doesn't matter. Along with its bright red distraction, we lust for it because of the detergentlike cleansing action the sugar in ketchup provides. Liquefied sugar is excellent at dissolving some of the fat sticking to the roof of the mouth. To get a bun that will be strong enough to hold all this together and still be attractively white inside, good old Plaster of Paris—the same material schoolchildren worldwide build with—is often added to the dough.

A big gulp of milk helps the bite go down. Most of the world would find this impossibly odd. The boys' very ability to digest milk marks them genetically. Only descendants of families from such areas as East Africa, the Arabian Peninsula, Europe, and parts of India—where livestock came to be used for providing milk—have this ability. Human physiology normally allows only babies to digest milk, and it was a lucky mutation, selected for in these areas and a few others, which allowed certain adults to maintain that ability.

Had the parents not been watching, the boys would have instead selected a soft drink. Carbon dioxide bubbles explode with delightful pain on tongues, and caffeine starts jolting into bloodstreams, and (especially) gobs and gobs of yet more sugar roll along in the general downpour. There are seven teaspoons in an average can of soda, which once would have made it impossibly expensive; before the Reformation sugar was a rare luxury crop. Honey was the more common sweetener. Monks produced a lot, keeping bees for the beeswax candles they needed for devotions. When monastic lands were sold off production dropped, at least locally; when slave empires began to appear, based on the harvesting of sugarcane in the tropics, supplies soared back up, and the spread of low-cost sugar began.

Caramel coloring turns the soft drink brown, which seems arbitrary, but isn't. Almost everyone judges a drink as sweeter than it is if it's colored brown. Orange juice makers get in on this notion by often using darkened glass, as did soft drink bottlers for years. But even colas served in clear plastic can still benefit from this perceptual flaw: Americans consume 126 pounds of sugar each year and without the caramel coloring to make those drinks seem sweeter than they are, even more sugar would probably have to be added.

The wife sits down, an impressive salad in hand. A century ago women were almost never allowed into restaurants, as it was considered far too unseemly. At an even earlier time, whoever was here would not have had the fork she's using. There aren't any forks in Leonardo's *Last Supper,* for the same reason that there aren't any modems in it: no one in Italian homes had any experience of them. Even as late as 1897 the Royal Navy had forbidden the use of forks as being "prejudicial to manliness."

An entire forkful of mixed vegetable fragments is lifted high, as one great mouth comes crashing down, crunching away untold billions of intricately evolved chlorophyll molecules. The first of the vegetables are propelled to the back of her throat, started on their long way down. Such vegetable intake is remarkably healthy for the eater. Vegetables are generally low-cal, for the excellent physiological reason that plants can always spread out their food reserves in tangles of roots, leaves, and spreading stems. No individual part—aside from the seeds—has to carry a dense quantity of nutrients, and so no individual part carries a lot of calories. Animals are the ones that carry their resources in the denser chunks we know as fat globules.

Vegetables are also good because they evolved in a world of direct, unblocked sunlight. Such sunlight breaks down cells, as people who've spent too long on the beach so wrinklingly demonstrate. The plants need some mechanism for building up those cells again, and for that they've evolved numerous substances, most notably the carotene molecule (which—the name is misleading—is common throughout the plant kingdom, and not just in carrots). The more a plant is exposed to sunlight, the more of this damage-fixing carotene there'll be, so when the wife swallows chunks of lettuce or spinach now, she's also swallowing carotene. It's rich in the powerful antioxidant beta-carotene. Human physiology is similar enough to plants—there are similar DNA and cell processes—that the carotene will perform a similar cell-rebuilding function in her.

The wife chews a tasty lettuce leaf next, which is about as low-cal as you can get—lettuce is 94 percent water—but still has carotene. How much depends on how dark it is. The general rule is that the more dark green chlorophyll in a leaf, the more carotene is clumped along with it. The deep green outer leaves of a lettuce can be fifty times higher than the pale inner leaves, which is why if she offers a forkful to the boys it would be foolish for them to refuse. The pathetically thin and pale lettuce fragments on their burgers have almost no carotene at all. Any dark green vegetable would be better. Broccoli is one of the best.

The next level down is peas that have been cooked from frozen. There are levels of quality to all vegetables, of course. Fresh peas are better than frozen, which are better than canned. The reason is that at harvesting time peas start metabolizing quickly. Metabolism generates heat, which destroys sugars, which creates starch, which tastes like glue. If peas are sent to a factory for freezing quickly enough you can avoid that. The peas that don't make it in time are the ones dumped into cans.

A salty chunk of chicken breast emerges from under a lettuce leaf. The chicken is easy to find, because it's so white. This is uncommon for food, which tends to be green or reddish or brown, but very rarely white. This chunk is an exception because of the sedentary lifestyle of the modern-day chicken. The flight muscles in the breast don't get used much, so there's no reason for oxygen-storing red blood cells to be soaked darkly through them. As a result, the breast comes out white. It's the same reason halibut and flounder and other deeply unadventurous hey-let's-stay-and-look-at-the-bottom-mud-some-more sea creatures also have white flesh. Only fast-swimming tuna—and fast-running chicken legs—get darker meat from all the energy-ready blood that has to be kept in place.

The bright chicken breast is also low-cal. This happens for excellent thermodynamic reasons. Chickens—and small animals generally—are so tiny that most of their bodies are quite close to the open air. They need to put what fat they have right up there, near that surface, to insulate against heat loss. Stripping off the skin means lifting off that blanket of fat. Will it be moist though? This is the great weakness of our otherwise thermodynamically perfect poultry, especially if it's stored for a long time, and so must be regularly fixed. An inspection of processing factories often reveals the surreal sight of definitely dead chickens lined up before banks of revivifying hypodermic needles—there can be six rows, with a hundred or more needles each—that tilt forward to mass inject saltwater into the chickens to keep them from drying later. Even without an inspection of the factory you can get a hint this happened by looking at the package a slice of chicken meat comes

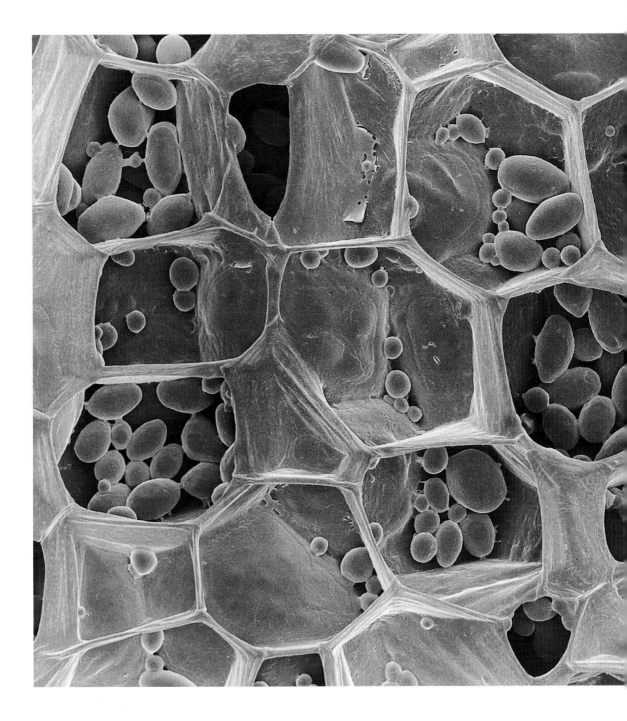

Heat a potato and the starch ovals swell like inflating footballs,
breaking down the otherwise indigestible cell walls.

in to see if the chemicals called polyphosphates are listed in the ingredients. It's a sure sign that you're paying for extra water. The hypodermics need to have some of the polyphosphates mixed in, to keep the saltwater from leaking out.

A curiously moist chicken piece gets swallowed, then the wife pauses, leaning back, to chat awhile with the boys. It's another one of those seemingly innocuous family habits, that can be of the greatest significance as it gets passed on. Overweight families, observed on video, hardly ever slow down their eating rate as they go through a meal. At Chinese restaurants in New York, average-weight diners reached for the chopsticks about 27 percent of the time in one study. Not the fat diners, though. They weren't going to be slowed down by anything: all but 2 percent grabbed for forks and spoons, to guarantee the all-crucial maximum intake speed. Their kids learn what's thought best.

Music comes out faster from the loudspeakers now, which doesn't help any efforts at moderation. Food court managers take a trick from the Muzak executives, and work carefully to speed their table-hogging customers away. In one Dallas eating area, families regularly went through their meals eleven minutes quicker when there was fast music on than when the music was slow. Listening to slow pieces of Brahms or Mozart does the reverse, which is why few except the most outrageously priced restaurants can afford it. With the more common fast music, the number of bites per minute goes up almost automatically, as hidden videos embarrassingly reveal. It's too irritating if the loudspeakers are pumping faster than 100 beats per minute—too many families then enter the food area, look around, and edge away—so as a compromise managers sometimes keep the music below 100 beats, but make sure the seats are uncomfortable. Benches with backs that are set too far away for easy leaning or are coated in slippery plastic, will do fine. On crowded weekend lunch hours managers can go further and raise the chilling air-conditioning: cool air invariably makes people feel like eating more.

The wife feels uncomfortable resting so long somehow and takes up her fork again; selecting some deep green spinach as a good way to keep her iron levels up. This isn't the best reason to eat spinach though, because it really isn't especially high in iron: the belief that it is comes from a misplaced decimal in a government report. Rather, this dangling spinach now being directed mouthward is excellent for vision, as indeed are all the other dark green vegetables. Carotene, besides repairing cells, also has the capacity to easily absorb light rays. Transferred to the omnivorous human, some is led, barely changed, to the retina of the eye. Within hours it's twisting away up there, helping trigger signals along the optic nerve to the brain.

It's too bad that no one is eating any garlic. Garlic is rich in allicin, which in the wild protects garlic plants against fungi and in the human body resists cold and flu viruses. It's capable, for example, of killing all the para-influenza 3 viruses it contacts. The onion in the salad has a related sulfurous compound. This has long been known for its effects on the mucus membranes of the eye (this is why we cry when peeling onions) but has recently been found to be equally powerful as a decongestant in the chest. These sulfur molecules are highly soluble, and the reason cutting an onion under running water helps. The irritant molecules merge with the water, and go down the drain. The chicken in the salad is rich in the amino acid cysteine, which in its variant acetylcysteine form is also good at thinning the fluids building up in the lungs of anyone suffering a cold. The old wives tale that chicken soup, rich in garlic and onions is good for your health is right.

The daughter arrives at the table, not in the best of moods, as her stomach is trying to digest itself. This is a problem all severe dieters are likely to have. The rest of this family, casually chomping down the food she craves, are lucky enough to be undergoing satisfying bouts of what is termed "retropulsion," which is the action of the stomach bottom sloshing and shearing their food. But her stomach is empty, and reduced to stripping off its own wall secretions to feed into the waiting intestine. Even this self-digestion doesn't help much. Everyone else around the table has also gradually begun slow wobblings of their intestinal muscles, as electrical signals start the intestine's toughened wall gently swinging inward and then outward, like a rope going up and down. But without having eaten any amount of food since dinner last night, she experiences painful contractions of much greater force; repeated once every three seconds for several minutes at a stretch before pausing, as the full twenty-five feet of her small intestine tries to drag along even the tiniest amounts of food that might be left in the stomach.

A dieter's torments are nearly impossible to avoid. Even the ordinary number of fat cells she's likely to have are lamentably dense. Toss a chunk of them on a fire, and they splutter madly, releasing their energy. When a teenager eats a little less, her body has no way of knowing that she's not in danger of starvation. It lowers its metabolism to make up for the change, and the dense fat-storage cells are barely affected. Only if she drops her food intake to staggeringly painful levels—overriding even the antistarvation defense—is there a chance of getting a noticeable number of them emptied.

She's also likely to poison herself, at least a little bit, if she really lowers her food levels. The dioxins everyone in the family ingested over the years from the butter and other rich sources they shared can't easily be expelled. Instead it has accumulated, locked away, fairly safely, in these out-of-the-way fat deposits. But let the teenager go on a strong enough diet, and her fat cells gush out everything they contain in a last-ditch effort to keep her from starving. The years-old dioxin comes along, reconnecting her with her family's past, as its levels rise noticeabley in her bloodstream now.

The thin plastic wrapping of a yogurt carton is torn back, and the suffering teen's spoon plunges in. If a handsome man is sitting nearby the spoon will probably just stay there: one Toronto study found that teenage girls almost always stop eating in that case, and will only pick up if the man leaves or gets distracted. (If it's a woman, and especially a chubby woman, who sits nearby, the spoon is delayed not at all.) Safely unobserved now, this daughter scoops her spoon through the container.

Since her diet allows only low-cal yogurt, she's in for a problem later. The sweet taste of artificial sweeteners makes the eater *think* she's taking in some sugar with each spoonful. The girl's blood glucose starts to go down by about 6 percent in anticipation of that change. But because no true glucose actually does come in, she'll end up stuck down there, with lower levels than originally, and before too long she'll be hungrier than when she started.

The whole idea of solving a weight problem through special diet foods reveals another curious gender difference in body perceptions. When the Pepsi company was promoting Diet Pepsi for the male market, they found that it didn't work to advertise it as having "no calories." Instead they had to

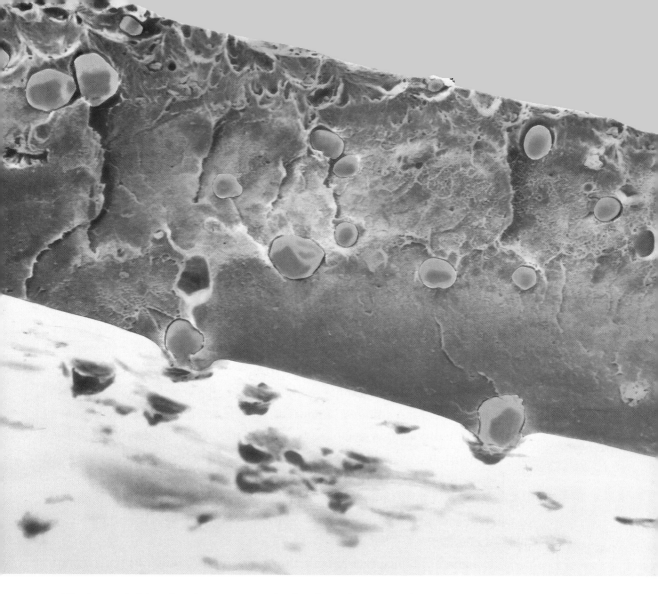

Biodegradable plastic, common in food packaging. The bright orange blobs are starch granules. Buried in the ground, the starch absorbs water and bursts the plastic open, exposing fragmented surfaces that bacteria can then digest more easily.

promote it as having "no sugar." When a man and a woman feel equally bad about their bodies, their surveys found, the man will usually assert that he just needs to exercise more; the woman, sticking to passive choices, will conclude that she needs to eat more diet food.

Some bran is sprinkled on the creamy useless yogurt. It's the sort of food dieters are convinced is good, because it certainly tastes bad enough. And although it's healthy in moderation, wheat bran is also dense with phytic acid, which can tie up much of the calcium in the yogurt. This is not especially good for anyone, let alone a nonexercising, sun-avoiding teenage female. Calcium is stiff and tough—the mortar holding the food court's bricks in place is made with it—and her osteoblast cells need a good two pounds of it so they can sculpt her bones into proper shape. Even a smaller amount available now could set her up for bone problems later. Oat bran would be better—it has hardly any of the calcium-grabbing phytic acid—and so would whole wheat bread: cooking destroys the undesired phytic acid, so only the stomach-filling safe bulk remains.

Many dieters rely on diet pills. They can work, but not in a way most users would appreciate. A typical diet pill simply mixes products that give a short-term water loss—caffeine and the irritant pesticide known as bunchu are typical here—with something really bulky, to give you the impression of being full. The bulk cattle feed known as cluster bean, grown in the southwestern United States, is commonly used for this. Once in the stomach it absorbs moisture, and—as with the stamp glues and cellulose pulp in baby-food formulas—inflates quickly in size, pressing hard against the stretch receptors there. A little bit of bran is sometimes included in the pills, although it will cost several dozen times more when consumed this way than when purchased directly.

The dieting teenager wouldn't think of drinking anything but plain water. In the waiting glass, a maelstrom of activity has been taking place. About one in every 10 million water molecules in the glass is constantly cracking apart, leaving a chunk of oxygen and hydrogen spinning off to one side, and a sliver of isolated hydrogen spinning off to the other. Without help the contents of the glass would soon disappear, in a quick-evaporating puff of gas. Luckily, elsewhere in the waiting glass a proportional number of earlier chunks are clasping back together frantically fast, to reconstitute water molecules. Dieters become dehydrated, which makes stretch receptors in the vein returning blood to her heart signal her brain strongly for fluids. Heeding the message in her brain, she drinks up.

The daughter tries to work out how many calories she's had so far. In a brain scan her frontal lobes would show up intense activity. The harder they have to work on these calculations, the more energy they will use. It doesn't help that she's trying to power up her brain despite having pretty much skipped breakfast. Starting the day without food can lower the level of the acetylcholine that carries key messages between brain cells and decreases your memory abilities hours later.

The dad's back, cheerfully carrying the mammoth-size pizza he's had to wait for. Lee Iacocca remembers, as a schoolboy in 1920s America, being beaten up when his classmates found that his mother had given him something as disgustingly foreign and un-American as pizza to eat. But so long as there's not too much fatty cheese it's an excellent food: balanced in vegetables and flour and rich in the same flooding antioxidants there were in the wife's salad and the morning's orange juice.

The family eats while clouds of carbon dioxide swirl from the cooking areas as well as from exhalations, all of it sucked into the mall's vents and sent tumbling outside.

5

separate meanderings

A half hour later and the meal's over; the older kids head out, and the parents settle back in their seats, finally alone; able to talk in quiet. Then the husband cracks his knuckles.

Cracking your knuckles doesn't sound horrible because you're violently grating human bone parts against each other. That would be a disgusting thing to do to yourself. The reason cracking your knuckles sounds horrible is because you're actually pulling human bone parts *away* from each other. Finger joints are not rigidly connected. There's lubricating fluid between them, and if you tug hard, especially between the metacarpal and the first phalanx, that fluid can't all escape from the gap. Instead, it now has to fill a larger volume. That lowers its pressure, and if a nice wrenching bone twist ensures that this happens quickly enough, then some of the water that's part of the fluid is instantly changed to a vapor state. X rays at this point show bubbles suddenly appearing. Like steam escaping from a boiling pot, they will fill up a much larger volume.

Perfect diatoms. Chunks of these marine algae help give face powder its distinctive smooth feel.

A tenth of a second or so into the abuse, it's all still quiet. Then the enlarged vapor bubbles push the wrenched joint so far apart that other fluid in the body, especially the fluid normally in place farther away around the joint, comes rushing in. It's blissful for the human being doing all the tugging and wrenching, but fatal for the vapor bubbles so produced. They can't survive against the inrushing fluid. Each individual bubble makes only a little crackle when it's crushed down, but since there can be very many bubbles the result is most audible.

A small number of air bubbles remain, keeping some of the enlarged distance. It takes about twenty minutes before the gas in them finally dissolves in the surrounding fluid, and the mighty vapor-collapse crunch can stir again.

At this point an argument might start, not necessarily about knuckle-cracking. Money, which TV programs to watch, and the kids are the most likely topics of argument in America, in that order. Once the quarrel begins, the process is surprisingly similar, regardless of the subject.

Even before any knuckle-cracking, often even before any words are exchanged at all, a curious physiological aura is passed back and forth between partners who are likely to argue. Husband and wife now enter a curious physiological state. Sitting near each other, merely knowing they're going to have to talk or share some time, their hearts start beating faster, blood pressure goes up, tiny geysers of sweat appear on their fingers, and even the speed of blood circulation in their fingertips suddenly rises. It usually can't be seen from outside, but social psychologists who've done the wiring up measurements have found that when it does occur—each partner inwardly just daring the other to start something—follow-up studies four years later show that these pretensed couples are the ones most likely to divorce. (In all cases, divorces are most likely in the third and fourth years of a marriage, when they occur at all.)

How to get in the blessed group of people who don't argue? A curiously deep division between good and bad relations was found in a study that interviewed a number of couples in-depth, and then came back four years later to see which ones were still together and happy about it, and which ones had divorced or were together but unhappy about it. In bad relations, the ones that ended up in divorce or grievously dissatisfied marriages four years later, if you asked the wife what bothered her about her partner she'd give the typical complaints which women, usually justifiably, have about men: she'd say that he didn't listen to her enough or that he didn't share his emotions or that he could be coarse in his habits. In good relations though, the ones where the couples stayed together and said they really enjoyed spending time with each other, the wife also might say yes, her husband didn't listen enough or share his emotions or he could be crude . . . but she didn't really mind. As the old saying has it, there are two marriages: his, and hers.

Barely half of American couples agree on whether their family is trying to keep a budget, and if they do agree, they often differ on which one of them is supposed to be doing it. Fifty percent of men say their wives vote the same way they do, but two-thirds of women say they voted differently. It gets worse.

Almost 75 percent of couples disagree on whether they've had a deep conversation about feelings in the past week, and about as many disagree on whether they've had an argument. Up to a third, if asked, will even—this is the sort of statistic Dave Barry likes—disagree on how frequently they've had sex that week. And nearly 1 percent, the truly befuddled, come up with different estimates about how many children they have (they may be the ones fighting over alimony and child support).

How can you tell which sort of relationship you're in? In the longer-lasting ones, both partners used the word "we" a lot when they told an interviewer those satisfyingly mythic stories about how they first met or how they decided to have kids or buy a house. But the other couples didn't: the past had been rewritten and they already were becoming two separate individuals. Even the past week was remembered differently. In a good relationship, each partner was likely to remember if the other one had done something nice—a friendly handhold, or shared laughter—recently. But in unions that were falling apart, even when one partner had done something like that recently, the other one would say that no, she had no memory of it at all. The arguments in these distressed couples can be disastrous for your health. Stress hormones pump into the bloodstream for a half hour or more after even minor run-ins, and when these are repeated enough, statistics show almost every disease rate going up. In couples that have stayed together a long time despite the arguments, women suffer most of all: the men must end up running on autopilot, for it's only the women's stress hormone levels that go up after a spat.

When marriages arranged solely for economic advantage were popular there would be fewer of these types of problems, because little emotional sharing was ever expected. When the young French aristocrat Alexis de Tocqueville visited America in the 1830s he was startled to find how offended American couples became if their partner wasn't friendly, as that was rarely a consideration in France. Romance was assumed to come from elsewhere, though if you had come to like your spouse, there might be some proprieties. One French lady informed the Seigneur of Brantôme that she always carefully assumed the top position when she was with her lover. That way, she explained, if her husband asked if she had ever let another man mount her, she could avoid hurting his feelings and quite honestly say that she hadn't.

A number of accurate predictions can be made about our modern relationships, long before the arguments and tension surges start. It helps, statistically, to have more of those background similarities we saw at breakfast: matches in age and education and looks and views. Sharing likes and dislikes in art turns out to be a good predictor of harmony; spending shared time drinking, though, is a good statistical predictor of the opposite. It also helps to not have broken up too many times before getting married. The couples that repeatedly dated, then broke up, then reconciled, in a repeated cycle of tears and hope before finally settling together, are the ones especially likely to divorce. If the harmonious life sounds boring, take heart: couples that don't feel the need to argue much have a far better sex life, enjoying two to three times the weekly average of those who do. This isn't that difficult, as the rate of those who argue is only about two times a week. (In all couples the rate is, statistically, controlled more by the wife's age than the husband's: the younger she is, the higher their average.) Combined with the fact that most Americans pray about five times a week, and many pray more often than that, this means, as the statistically deft sociologist Father Andrew Greeley has pointed out, that the average mall shopper you pass in America is more likely to have prayed that day than to have had sex.

The baby wakes up and looks around. This continued restlessness—why are the afternoon naps disappearing, but only on weekends?—puzzles the parents. The problem is the coffee that the mother drank at breakfast. Once it dissolved in her bloodstream there was nothing to stop it from slipping into her blood and then into her breast milk. This means that the baby will have received a whopping dose at its feeding just before lunch.

To make things worse, not every individual in this family can get rid of caffeine at the same rate. If the family dog had accidentally lapped some up from an abandoned coffee cup nobody would have to worry, for dogs have oxygen-insertion mechanisms powerful enough to deactivate caffeine about twice as quickly as adult humans. In two hours about half of what it might have absorbed would be gone, compared to the five hours or more for the grown-ups. Ten year olds are halfway between beagles and adults, so if the brother had sneaked some coffee at breakfast, he'd be over its effects by now, too. A baby, however, is all the way on the other side, taking seven hours to clear just half of what they intake, and so will have notable levels surviving even now. The younger the baby, the stronger the effect, which is why coffee drinking is not recommended to nursing parents who have any interest in sleep. Fetuses are even more inefficient, and whatever amount they get dumped with, crossing from the mother's bloodstream, largely stays put. By one study, most babies born in America come out with measurable caffeine levels in their blood. The average is about 1.5 to 2.0 micrograms of caffeine per milliliter of umbilical cord blood at birth, and it's two to three weeks before it's all gone.

The boy and his friend are out exploring in the mall by now. The parents were willing to let them wander unsupervised because they trust in the protecting gaze of closed-circuit cameras and security guards. They're still not as safe though as a random walker in a capital city early in this century would have been. When the young Peter Medawar was preparing to go to England for boarding school, in the 1920s, his parents told him that when he got off at Waterloo station in south London he should find some stranger, tell them he was a young schoolboy, new in the city, and ask if they could please tell him how to find his way to King's Cross station in north London. Then he was to cross the city by himself. Medawar of course had no problem, as all the strangers he encountered were reliable.

The teenage girl meanwhile has joined some friends, and they have clustered outside the jeans store. The girls are ostensibly talking to each other, but since they all know that they're just waiting for the guys to show up, they also know that no one's really listening. Instead, they're doing what teenage girls seemingly have an irrepressible reflex to do, which is, in a rush of scattered glances and dead-solid appraisals, judge how each other is dressed, and who's most popular, and so estimate who's going to have the best shot with the guys.

Bodies need to be adorned for this task, whence all the polished rocks or metal earrings, the crushed silicas and other boulder dust as foundation powder on faces. That's not to mention the dead cow skin and hooves ground

Opposite: X ray of a young woman standing up, with the colors highlighting the stress pattern in her skull. The necklace isn't floating, but resting on her neck.

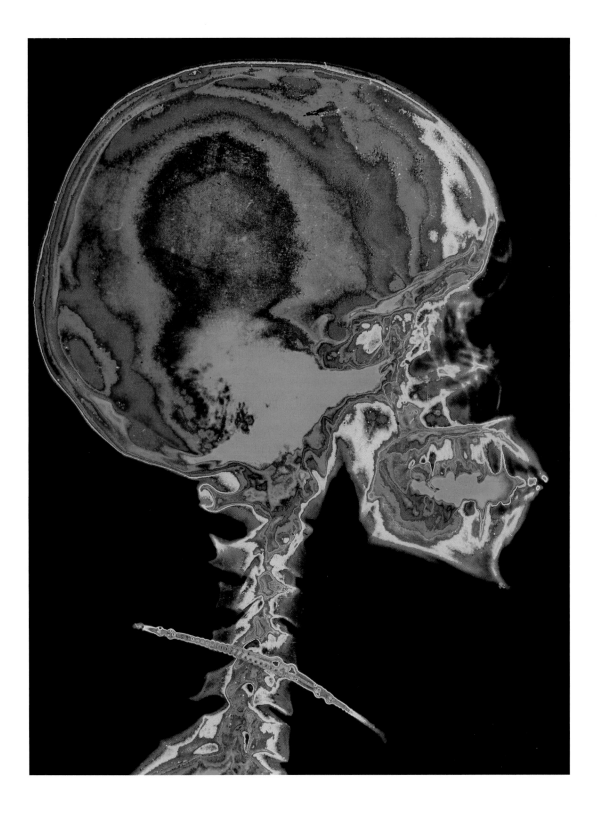

up in the polish on fingertips, and the even brighter glow in the lipstick. Cheaper brands of lipstick get their luster from mere chemical factory products, but more expensive brands get it from authentic fish scales, scraped from the eviscerated creatures and then processed at the cosmetics factories. Sometimes people have gone too far in competitive adornment. Queen Elizabeth I used white lead paint to make her face attractive. Unfortunately, lead is a poison, so it steadily ate into her flesh, and she needed to apply more and more layers before going out. In time she was using wool pads stuffed in her cheeks to mask her lost teeth, the desperately thick layers of white lead on her face, egg white over the white lead to cover the skin fissures that still showed through, and blue paint on the very top, to trace what could (if you were a courtier and your income depended on saying really nice things) be charitably described as suggestive of a young beauty's fine blue veins. Being queen though, the effects were not as distressing as they would be to others: she simply banned all mirrors from her palace so she wouldn't have to see the consequences.

Parents say that teenagers' hormones are raging, but actually they're pulsed out only a few times each day, in fast surges that last barely three to five minutes each. Stress can change the pattern. This is especially important in the autumn, when bone growth is up to two and a half times faster than at other times. There can be a vicious cycle, where after missing out on one season of this bone growth, a developing teen feels so stressed by it that her body keeps on blocking or reducing the steroid pulses that would let her catch up.

Along with the competitions though, there will also be subtle linkages among the teens. When women spontaneously tell stories about themselves, they often describe group action as succeeding; in men's stories, by contrast, it's usually a single hero, acting alone, who triumphs. There are also chemical linkages in a female group. Humans generate a tremendous number of lightweight chemicals, as we saw with the parents' immune clouds, and these steadily float loose into the surrounding air. Pregnant chemists have to be careful when they work near certain sensitive equipment, because the altered levels of steroids they're exhaling can change its calibration. Something similar seems to be the reason that girls who spend a lot of time together—school classes will sometimes be enough, though sharing a dorm is even better—

sometimes synchronize their monthly cycles. The exact chemicals being evaporated are as yet undetected, though the evidence is persuasive.

The teen girls are also likely to be keeping their voices at a precisely shared level. They take their cues, in large part, simply from how bright the light is. In one California college, turning on only a third of the fluorescent lights in a popular corridor kept the conversations down to an easy fifty decibels. With the rest of the overhead lights on, conversation rose to sixty decibels.

Racial tensions aren't likely to be much of a problem. To visiting foreigners this has been the most extreme change in America in the past generation: about 85 percent of American teens now say that they have one good friend of a different skin color . . . though they quickly go on to say that most members of their own ethnic group are more prejudiced. In Britain about half of all dark-skinned citizens originally from the Caribbean marry white Britons. By the late 1990s the number of children born to parents of different races in the U.S., though still quite low, was growing faster than the number born to same-race parents.

The guys finally arrive, and now even the pretense of female solidarity disappears. Seemingly intelligent female humans break off all conversation with each other, just so they can turn to the boys, these paragons of all satisfaction, and focus on them and smile and generally simper into their affections. Girls who had been standing straight will often start tilting their heads slightly to the side, in the universal posture of submissive interest (which also crops up in girls' high school yearbook pictures far more than it does in boys'). Occasionally there seems a violation of the simpering and neck vertebrae shifts, as when two girls suddenly start talking louder, ostensibly concentrating on each other, pretending that they're really not concerned at all with what's going on. But look closely, and they'll usually be glancing around to see that they're getting the right effect.

The boys return this attention in young male style. They grunt and laugh too loudly and interrupt each other and certainly interrupt the girls. Despite this apparent confidence, all but the closest of male friends are likely to mistrust each other, too, at this age. Teenage boys rarely look at each other's eyes as much as girls do. They stand farther from each other when they talk—which we also saw with the dad and the package deliverer in the morning—and they don't give as many of the little supportive nods and grunts that girls do. It used to be thought that stress was worst for the least popular boys. But studies of African wild dogs have shown that the dominant males in hunting packs regularly have more debilitating stress hormones in their blood than the others. It's exhausting to try to stay on top, and tension between young males of our species isn't new. In a section of the medieval Fishergate cemetery in York, spanning the eleventh to the twelfth century, twenty-nine bodies had bone injuries made by sharp instruments. Most were "injuries from edged weapons using a slicing or thrusting action, or consistent with penetration by an arrow or a crossbow bolt. . . . Some had been crippled by having their thighs hacked, some had died by thrusts in the back while lying down, some had been decapitated. . . . The victims were mainly young males."

What's really going on here is practice in selecting a mate. In the coldest of chemical views, our daughter and her friends are responding to their strong impulses to get the hydrogen bonds of one long molecule inside of them—their DNA—configured with the hydrogen bonds of the most desirable male DNA available. As they can't stick a miniature periscope inside to check out the arrangement of hydrogen bonds directly, adolescents are consigned to using indirect signs, whence the frantic attempts to give hints of maximum wealth, looks, and esteem. The rankings are surprisingly similar around the world: when thirty-seven different cultures were analyzed, women always said they were keeping men's earning power in mind more often than almost any other factor when deciding whom to settle with. Men, however, weren't so concerned with women's incomes, but instead ranked the best indirect sign of fertility—their looks—as more important.

To help matters along, the teens have powerful sex hormones floating in their bloodstreams. The boys' testosterone levels have been climbing or crashing all through the day. There are six or seven peaks daily, with the highest levels in the morning, and the lowest levels about 25 percent down, in the evenings. Even at its lowest point, though, there is plenty of testosterone for sexual purposes. Women always have a small amount of male sex

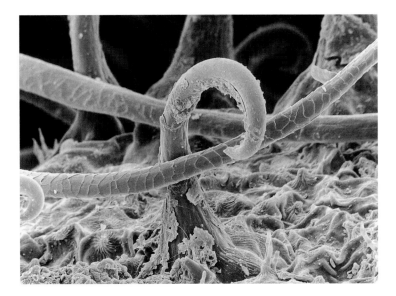

In 1948 a Swiss engineer examining burrs (*above*) which had caught on his socks came up with the idea of an artificial fastener where tough plastic hooks would grab slippery loops—what we now call "Velcro."

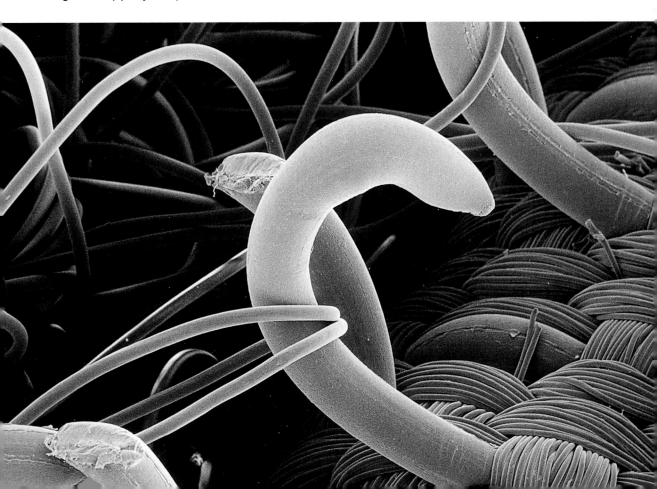

hormones in their blood, just as men have a similarly small amount of the female hormone estrogen. Among teenage girls, sexual interest generally reaches its peak just when these slight testosterone levels are at their highest—an effect that occurs, with convenient evolutionary sense, right about the times of maximum monthly fertility.

The daughter tries to keep a fixed smile on her face, as she longingly admires how the others manage to blend in, and also tries to work out which guy in particular—the shy one near the back?—she's supposed to meet. She's also trying to decide whether to remove her eyeglasses. Taking them off might seem the thing to do since girls who do that are regularly judged as more attractive than ones who leave their glasses on (an effect which, unfairly, barely occurs when boys are evaluated). But who wants only to seem more attractive? Girls who take them off are also believed dumber than ones who leave them on. Those lenses kept firmly in place mean a teenage girl's IQ will be thought to be twelve points higher than otherwise. It's an effect that lasts only so long as she stays quiet, though. The moment someone talks, she is judged by what she says, and the effect of wearing glasses vanishes. That fixed smile can hurt, too. Women are often perceived less intelligent if they smile *too* much.

First names also affect popularity. Among boys in the United States, the four most hated names in the latest survey are Myron, Reginald, Edmund, and, surpassing them all so much that it's hardly ever used anymore, Oswald. The most popular first name in America, appearing on top in surveys year after year, is (ahem), David.

Eager to look ultracool, the daughter reaches into her pocket and extracts a single long cylinder of plant fibers. To get the nicotine from that cylinder into her body isn't a simple process. First a match has to be struck—one of several hundred million in America each day. As smoking isn't allowed, it'll have to be furtive. The flame explodes, then, applied to the end of the cigarette, it makes the raw paper melt, then quickly vaporize. That ignites the densely packed tobacco inside. When a studiously casual inhalation from the teenager suctions many trillions of free-floating oxygen molecules to that tip, a great bellows roar raises the flame's temperature ferociously hotter, and that finally is enough to vaporize some of the nicotine molecules locked even deeper within the cigarette.

It also, unfortunately, is hot enough to create the tiny molecules called polycyclic hydrocarbons. Those ride into her lungs unnoticed, stuck on the top of each smoke particle like spots on a beachball. While the nicotine travels on to the brain, the polycyclic hydrocarbons fall off where they land, with the smoke of course damaging sensitive throat-lining cells as it swirls past. Teenage boys are likely to be smoking unfiltered cigarettes, and so the smoke particles are so heavy that they only reach the higher parts of the airways before the polycyclic hydrocarbons tumble off. The daughter, although aspiring to this street-tough look, too, is more likely to have a filtered brand, that lets its smaller particles go down, down, down to the very inner cavities of the lungs before dumping the polycyclics. This is why women smokers often get their first tumors deep in their lungs; male smokers usually get them higher up.

The girl coughs once, moving some of the debris along, but not really helping much. The amoebalike macrophage families in her lungs have to start hauling themselves forward, to try getting rid of the rest of the stuff. But the polycyclics are more than a match for the macrophages. If the girl stops smoking on her nineteenth birthday, she'll still be clearing her lungs of leftover polycyclic debris when she's almost thirty.

Should she be doing this at all? Everyone says not to smoke, but nicotine works on the limbic system cells in the brain, which can give the most confidence-inspiring emotional signal when properly stimulated. Teenage girls especially need that, which is why ads aimed at them work so well.

Tobacco comes loaded with nicotine, simply another alkaloid pesticide that plants evolved to defend against insects. Nicotine slows the stomach's usual movements. Dieting teenagers love this. With food staying in the stomach longer, they feel fuller, and really do have less of an appetite.

The girl looks up, for suddenly one member of the student group—the leader of the girls—has rotated a bony hinge at the side of her face, and, jaw opening wide, begun to vibrate her vocal cords in a strange fast rhythm, while simultaneously expelling huge air gulps. This has a powerfully addictive effect, and almost instantly everyone else joins in this laughter, including, a little nervously, the daughter. But she missed whatever it was that caused the laughter, and so she's feeling a little uncomfortable. Could someone have made a joke about her? She blinks, then blinks again, and turns her head. She hopes no one will see, but she's crying now.

Father and baby are sitting together on a bench in another part of the busy mall, studiously analyzing the buoyancy properties of a shiny oblate spheroid. To the baby it's truly baffling: he leans close to it but the father tugs down on the string and it suddenly disappears. Just as quickly—and there are delighted giggles now—this balloon will pop right back up. It's hard to get items that float in air this way. In a denser fluid it would be easy, and even a rock on a string would float overhead, which indeed is how the continent the father and boy are perched on keeps from sinking down to the center of the earth: it floats on the even denser magma below.

It's time for the next stage of their mall afternoon ritual, which is the sharing of the ice cream cone. This means the dad buying one from the stand beside the nearby fountain, which is messy enough while holding a stretching baby, but then he has to sit down and share it, which can be even sloppier. Strategically placed paper napkins help a little, but babies are superb sugar-hunting machines, and once they've had an initial lick of the ice cream, they want *more!*

The reason babies are so clumsy is that their brains are unfinished. The nerve fibers inside adult brains are almost always wrapped in slender jackets of fatty insulation. With these jackets in place, nerve signals travel fast, and accurate distance-judging and hand-controlling systems are built up. Babies don't have as much of the fat-dense nerve cell insulation in place yet—one reason they're so eager for the wobblingly attractive ice cream, with its thick emulsifiers and margarinelike fats, part of which will end up wrapped around those brain cells a few hours after swallowing. Sticky cholesterol helps form these nerve-wrapping insulations, with the curious result that our brain is in large part made of cholesterol.

The dad's a little tired, and yawns now. Fat from the cheese on his pizza has triggered the release of amino acids in his gut, and one of these, cholecystokinin, is especially noted for its power to increase drowsiness—a powerful cause of our general postlunch lassitude. But the baby reasserts his control by hurriedly making its tiny lungs puff out a word-resembling sound now. That brings the dad back to full alertness. In response, without any feeling of demeaning himself, he immediately contorts his chest wall and especially his throat and laryngeal muscles to pump out quick, high-frequency falsetto squeaks—baby talk.

Babies can't hear deep sounds well, partly because of their tiny ear bones, and with their incomplete brain cabling they certainly can't process the subtle shifts in sound we make in quick multisyllable talk. So they manipulate us by responding to high-frequency sounds. Babies haven't developed this skill only recently: twenty centuries ago, Roman grammarians remarked on the strange adult squeaks babies induce.

The dad could spend all day, cozy in this shared world of voice and touch. But then he sees some teenage girls whom he recognizes from his daughter's school and smiles at them. They register his attention about as much as they do the air-conditioning ducts overhead. He checks his reflection in a shop window, wondering if he really is that haggard, and then, despite his joyful child—indeed, now exacerbated by the contrast with the perfect fresh skin of his child—he begins to suffer, once again, the parent's eternal lament.

He's feeling old.

There's good reason for this, as humans seem built to hold together properly for just about long enough to produce and raise children. Decline usually starts at about twice the age of puberty, which makes the onset of middle age soon after thirty, though the exact moment has been the subject of much debate. John Donne thought the ideal age to be reincarnated at was thirty, while Aristotle pushed it a bit more and said that thirty-seven was the healthiest age. (He pushed even further and held out for age fifty-seven as the overall peak—he didn't have to wear hip-hugging blue jeans, so he could say that—on the grounds that life's wisdom should be taken into account.) Some observers are more cynical. John von Neumann, the great mathematician, observed that middle age is generally defined as occuring at age x+4, where x is the worrier's current age. A less kindly definition is that simply pondering the matter is the sign that you've arrived at middle age, for when has a truly young person worried about it?

Many of the changes are only too noticeable, such as a receding hairline or baldness in men. There's also the matter of the slightly too prominent pouches under his eyes, where saltwater-filled bags dangle in notably useless storage. (This morning's argument with his daughter might also have made that worse, by disturbing his output of vasopressin hormone, which normally helps keep that under some control.) And then there's the giveaway eye-corner wrinkles, where the connective tissue under his skin is bunching up, and all the rest. Dictionary editors might be consoled that vocabulary goes up with age, at least for college graduates, from 22,000 words at age twenty-two, to 40,000 or more by age sixty-five, but most individuals would probably be willing to trade the comprehension of abdominal isoclines (parallel gradations of stomach fat) for the satisfaction of not possessing them. It's true that the great curved mass of the omentum—the fatty sheet trundled around on the abdomen—actually contains active parts of the immune system. But even slender individuals have enough, and those who've given up on sit-ups are merely spreading out those immune units, not increasing them.

It doesn't help that this is also the period when career possibilities begin to close down. A poll of personnel officers in the London financial world found that job applicants over thirty were going to have a difficult time, while anyone over forty could just as usefully make a paper airplane out of his résumé. Being locked in your current job wouldn't be so bad if you were happy with it, but that's rarely so. Mathematicians like their careers—over 90 percent would choose them again—but they are the exception. Most professional jobs are so disliked that rarely will more than 50 percent of the people doing the work say, if interviewed in private, that they'd willingly choose the same career again. Blue collar workers rank their satisfaction with their jobs even lower. The problem, in part, is that the western notion of career goals came strongly from the religious notion of predestination, and that's a lot to load onto a decision you made on a hunch or by chance or just by simple inertia, years and years before.

If the dad wants to feel really bad, he could reflect on all the subsurface collapses he can't yet see. He's shrinking about a quarter inch by this time in the afternoon each day because the discs between his spine bones are getting squashed. For men, from age forty-five there's significantly less testosterone produced, due to reduced blood flow to the testes; for women, the production of healthy egg cells also usually diminishes by that age, with its accompanying hormonal shifts. His eyesight is probably worse than before, for cells within the lens of the eye can't have many blood vessels, as that would block our vision. Dead cells within that lens have no way to get out, so when they

Inside our eyes, dead cells stack like ziggurats to form the transparent lens. Magnification: 3,000 times.

die they stay there. Through the years their dense, useless bodies stack up, making the lens increasingly stiff and hard to focus, which is one reason people become farsighted. It doesn't help that eyeballs grow unevenly, often ending up too long from front to back. You need strong yanks from the tiny ciliary muscles to squash the stiffened wriggling glob back into shape. If someone's muscles have weakened with age, he's sunk.

His brain has begun shrinking in weight recently, too, and is now likely to have less of the full three pounds weight it had in his early twenties. And, perhaps most disturbingly of all, the very brain cells with which we're supposed to supervise these impermanent assets, are themselves increasingly popping out of existence. This raises the question of where our bits of disappearing personality and memory have crashed away to. Men especially try to avoid knowledge of such facts. The general medical knowledge of women is so much better that not only do they know more about their bodies, but they also usually know more about men's bodies: more women than men, for example, can locate the prostate gland, even though only men possess it. But avoidance is of no avail.

All our body cells have a timing mechanism, set in motion when they're first formed, that accurately determines their life span. Sometimes it's very short: the cells on the inside of the mouth last only a few days before dying and slipping off for a final flotation funeral in our saliva, down this internal Ganges to the outlet in the all-consuming stomach sea. Some of the cells are set for a little longer: the liver cells created in the baby this afternoon, from energy supplied by the salad and pizza slices, will survive inside his body till he's four or five before dying and being replaced. Memory B's, as we saw, last even longer.

Brain cells, however, are never replaced, which means that the ones you have now are some of the original 100 billion or so you were born with. They've been firing away in the wet chemical battery of your brain all that time, sloshing amid the reactive oxygen and warm water, which might not seem the ideal place to store a great number of such fragile entities. Now, one after another, they're failing. By age forty, an estimated 5 percent of the brain cells in the networks installing memories will start dying off each decade. These micro brain deaths are never a sudden flash of electricity. Instead, like a computer dimming as its battery runs down, the brain cell that is flickering out of existence at this moment took several hours to reach its end.

It sounds terrible, but a little arithmetic is reassuring. If 500 brain cells leak out of existence inside our heads each hour—an estimate that has been much debated—it adds up to 12,000 a day, or 4.4 million a year. That's 176 million in forty years, which sounds like a lot, but a ten-month-old baby has nearly 100 billion to start with, as we all did at that early age. The amount you lose in forty years then is well under $\frac{1}{4}$ of 1 percent of that total. If anything, getting stuck in stodgy thinking patterns with the remaining 99.75 percent of brain cells is the problem. But look at the way the baby can cable up brain circuitry in fresh patterns. Adults can too, albeit with more effort—which is just what a return to full-out play with the waiting baby, eager to have us concentrate on comprehending its wobbling gestures, will help produce now. Even the preliminary step of loosening one's necktie can sometimes help. It lowers arterial pressure, and in a number of lucky people, will reduce the eyeball distortion that blurs vision.

The wife is wandering elsewhere in the mall, untroubled by any of these biological changes. In fact she's uniquely contented, at this, the only truly free interval of her week; the one time when she's neither commuting nor working nor paying bills nor (however much she loves them) having to talk with the kids or her husband. She can do exactly what she wants. There might be a certain tinge of guilt—the universal lot of modern working women—but for a few hours, once a week, she can handle that. It's even preferred: only 15 percent of American women, in a survey by the Factory Outlet Marketing Association, wanted their husbands to come with them when they went shopping for clothes. Men are less confident when it comes to clothes hunting, and over 35 percent want their wives along when they shop. One reason for the wife's independence is the immense pleasure that comes from a successful hunt: from the same marketing association survey, 36 percent of the women polled said they would prefer finding a great clothing bargain to having great sex.

But she has a more pressing need to deal with, and looks for a sign for the ladies' room. On a warm day there's time for a casual scanning. When it's cooler, the need is more pressing. Such weather changes make your blood volume slowly decrease, and send more liquid from the blood into the kidneys and ultimately the bladder. Mall managers know that on the first days of a cold spell, the mall's rest rooms will be particularly busy. Women especially suffer, for their bladders are slightly smaller than men's (even taking into account their lower average body sizes) because of the extra space the uterus takes up. Today there's likely not to be a line, and afterward the wife won't be crowded when she checks her hair in front of the mirror before going out. This is where honesty is truly gauged, for brutally unromantic social psychologists have done some timing with stopwatches here. If someone's on a first or early date, she is likely to spend about 58 seconds grooming here. But if she is married? The average plummets to a mere 9.8 seconds.

She'll have washed her hands now, but most likely done a crummy job of it. Even if you lather on the soap and pride yourself on really rubbing your hands together, you're likely to make the same mistakes that were noticed in a now-famous surreptitious study of the handwashing techniques of Australian doctors in a neonatal unit. Right-handed doctors almost never cleaned the inside corner of their left thumb, left-handed ones missed the inside corner of the right thumbs; almost none of the doctors or nurses, no matter how often they were prompted, could remember to do the *tops* of their fingers. This is especially grievous, for it misses the snugly protected area under the fingernails, where an ecologically formidable biota comes to reside; it also misses the exposed fingertips, where samples of virtually everything that's been touched during the day will end up. (It's hard to get people to believe that they've ever been at fault here. When results of the study were announced, few doctors in the unit admitted surprise. They themselves had known they were skillful washers, almost all said it was just the other ones who were lax.)

In the first shop there's such a richness of items available that once again the wife will be led by the needs of the shop owner. There's that reflex to veer toward the right, as we've seen at the mall entrance, which is why even inside a store the most expensive items are likely to be stacked on the right-hand aisles. It's not good to let shoppers walk through the store without something to carry their selections in, for people almost always stop shopping once they've reached the limit of what they can carry. Large items are usually set out as close to the entrance as possible, so if hand baskets are available they'll be taken; in a supermarket it goes even further, and bulky food items are near the entrance so shopping carts will be selected instead of baskets.

Properly equipped, and strolling along that oddly beckoning right-hand path, a curious thing happens to how fast our wife walks. When an aisle is narrow we're likely to speed up; if it's wide we slow down. More expensive items, will oh, just by chance, end up stacked on the wider aisles where she's going to linger. They will not be stacked in just any arrangement. Most people try to scan widely, but their first focus is going to be on what's conveniently at eye level. It'll also most likely be at the middle or outer ends of aisles—what are termed the "hot spots." Items often sell twice as quickly here as elsewhere. The owners know that, and right there, most easy to notice, examine, take down, and touch and feel; this is where the items with the highest profit margins go.

Background music is a little different in this store than it was at the food court or in the elevator. The goal is slow meandering, and so, with the exception of high-gloss teenage boutiques, tunes with slower tempos are almost always chosen. Seventy beats per minute—heartbeat rate—work the best, but it's not so easy getting the chord mix right. Happy music with major chords is what customers like to listen to, but sadder music with minor chords make sales go up more. The result is likely to be a mix. Blink rates start sinking as the wistful music and mazelike aisles take effect, and soon the shopper is likely to be in that blissful autopilot realm of a hypnotic session beginning, where the great sudsy eyelid sweeps are only happening at twenty times a second or less—the level the husband sank to in his couch-dwelling TV session.

In that vulnerable, prelogical state, colors begin to have a stronger effect. Bright ones such as red or stark yellow on packages or shelf edgings, are likely to start raising blood pressure and make skin pores push out slightly more of their salt-rich perspiration water, as physiological measurements of volunteer consumers have shown. (This is why you rarely see yellow phone booths: people finish their calls too quickly.) Blue is likely to have the opposite effect from red, and shoppers, interrupted from their glaze-eyed wanderings, even seemingly intelligent ones, will often say they chose a dark blue packet because it's more trustworthy, solid, and worth more.

The result is that people select intuitively, quickly. Most items were scanned for only one-fifth of a second, and even the ones that are selected seldom receive more than twelve seconds of evaluation. Shoppers also are usually pathetically inaccurate in judging the prices of what they do take.

Hardly anyone—young shoppers are the worst here—compares what he's taking with the prices of alternative brands nearby. About 25 percent of shoppers never check any prices at all. (This is especially likely with men, particularly men shopping for clothes.)

At least 30 percent of adult women in America are still using the same brands of major cosmetics they started with as teens, and almost 7 million are using the identical brand of mascara. The continuity is so strong, and passed along in families so closely—despite their protests of individuality, daughters tend to share their mothers' specific brand preferences, especially in cosmetics—that over two dozen of the top brands from the 1930s are still selling well today.

Labels rarely have enough information on them to allow shoppers to recognize the similarities. In addition, studies show that few people read the labels for content. People simply notice how long it is, and the more information listed, the more likely the item is to be bought. Shoppers over fifty were likely to read a label all the way through so they could make their own judgments.

Shoppers who want to cut through all the puffery do have one final hope. It only applies to more expensive items—appliances and stereos and furniture and the like—but it's a good one. Labels can mislead and packaging can lie but a warranty can't. The company's actuaries insist on that, for otherwise the company could go broke. This, accordingly, is the one item marketing executives themselves often look for when they shop. It's the one statement which has to give honest an assessment of how well a product is made.

The wife moves along to the clothing section. Items the shop especially wants to sell—the sweaters or jeans or blouses with the highest profit—are likely to be laid out on large tables. It's wasteful in terms of space but it performs the all-important task of encouraging "petting" of the fabrics. People are far more likely to grip and fondle items that are out on a table. And once they touch the articles, it's hard to avoid trying them on.

Alone in her cubicle with the new jeans, she struggles to get them on.

This is usually hardest in the autumn or winter because everyone gets fatter then, usually by four pounds or more—an effect that's so regular, that airlines take it into account in making accurate fuel estimates. The wife goes to the clerk at the cash register to pay for her purchase. The idea of having an automatic cash register is an American one, developed not so much to ease matters for the consumer but more to let the owner be sure that the sales staff isn't skimming his takings.

After buying the jeans that she has succeeded in squeezing into, the wife treats herself to a small piece of chocolate. Instantly she feels better. No one knows exactly why chocolate feels this good. Certainly it has a lot of phenylethylamine in it, which is a subtle brain stimulant, but cauliflowers are also packed with phenylethylamine, and are rarely sought in moments of acute postchanging-room distress. Chocolate also has some of the chemicals found in marijuana, but probably at levels too low to have much effect. What is unique to chocolate, however, is its ability to change chemical states very quickly. This, rather than crude psychology, must be the explanation. Lifted from its cool package, a solitary chocolate segment remains haughty, aloof. Even when the tongue pokes at it in its first urgent, demanding caresses, chocolate resists, with no outer sign of change at all. But let the steamy 98-degree warm tongue lap over it a second time, and the chocolate can resist no longer: a sudden liquid flood pours off as the entire upper layer melts from within. The tongue is briefly cooled by the chemistry of this phase transition, but given the ardor—the sheer, animalistic desire—of the determined chocolate lover, the tongue soon warms again, and with that the remaining chocolate is ready to be taken, melting in a sudden flood once more.

If this impersonal chemistry fails as an explanation, there is also the matter of the theobromine in chocolate, which can lead to increased hormone secretions, and, of course, the sugar. Ordinary chocolates can be one-third or one-half straight sugar, and the sudden high from this blasting through the bloodstream, even aside from the manner of its ladling in by the tongue, is bound to have some effect. A certain fraction of the population metabolize the sugar of a true chocolate binge so efficiently that it raises their blood alcohol levels, too. It's called the "auto-brewery syndrome," and depends on the particular mix of yeast species and bacteria resident in the consumer's

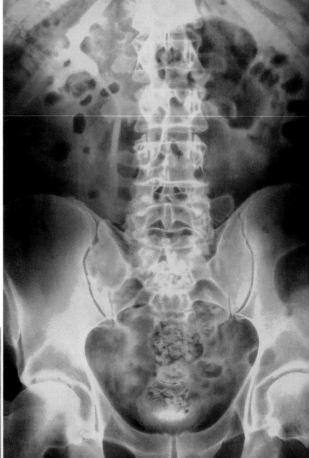

Agony and the inland sea: a bladder
when full *(below)* and empty *(right)*.

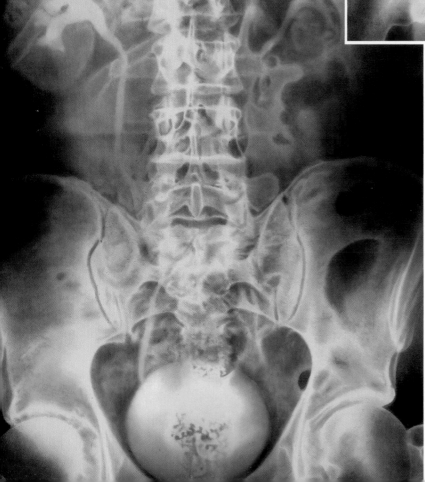

intestines. In one case, a joy-inducing level of 20 mg alcohol per 100 ml of blood was produced this way. Unfortunately, this took place in Sweden, where driving laws are very strict, and led to one utterly startled chocolate junkie being ticketed for drunk driving.

Adults are even less truthful about their chocolate consumption than they are about vegetable eating. Garbagologists in Green Valley, Arizona, asked people how much chocolate they ate, and once night fell, undertook their true investigative role and looked into garbage cans to see how true the answers were. People hadn't underestimated their true count by a mere 10 percent, or so. In a society besieged by food puritans, who would dare to reveal the awful truth? The actual consumption was twenty times—2,000 percent—higher than admitted.

The boys meanwhile are in the darkened cool movie theater, undergoing agony. The popcorn they're sharing comes loaded with salt, and this draws water from the rest of the body into the bloodstream. That raises the volume of their blood, making their bodies swell slightly, Michelin-man fashion, and their blood pressure goes up. But it doesn't last long, for the kidneys get to work, pumping the excess water back out. Only this time, the fluid transfer can't go gently back to be dispersed throughout the whole body. Instead it is squirted into the waiting bladder, which begins to swell, and get rounder, and then, really bloated, triggering stretch receptors on its inner layers. Those send various awareness messages to the brain. First a small number of signals, what will be interpreted as UM, IF YOU GET A CHANCE messages, and which are still pretty easy to ignore; then, as the circumference gets pumped groaningly wider, more numerous signals, interpreted as **I STRONGLY SUGGEST YOU DO SOMETHING!!** messages to the brain. At one time it would have been made worse by the power of suggestion from water streams constantly appearing before the boys. Modern movie screens have tiny angled prisms on them to rebound light, but early ones didn't. Liquids often reflect light very well however, and patrons of early movies had to sit before a screen that was kept brightly wet by tiny streams or constant mists of water running down it. Current movie houses can be darker, with only the red glow of the emergency exit signs visible beside the screen. You can't have ordinary light sources in the signs, because they'd fail if the electricity supply went out in a fire. Instead, an indignity greater even than that of the breakfast smoke detector is suffered, for in those signs, steadily roar-decaying for the boys to ignore, there are sealed volumes of radioactive tritium—modified hydrogen atoms dating back to the universe's big bang creation itself.

Kids can take the present-day water pressure agony longer than their parents, partly because their bladder tissues are more malleable, but also because there are so many great things going on in the darkened movie house that would be terrible to miss. There's food, to begin with, enough to make the mother's single chocolate insignificant: a range of candy bars and jelly beans and caramels of course, and maybe those little peanut-butter buttons, but definitely, almost defining the movie experience, there's going to be popcorn. Aztecs enjoyed it, though without flickering screen images in accompaniment, for they too had access to corn with water residues inside that when heated, would explode out as steam, distorting the kernel into this scrumptious snack. Modern popcorn is a wonderful invention for people who make bathroom scales, as the butter dripping over it is not, usually, made of anything that came from churned milk. Often it's the far more glutinous coconut oil instead, a substance that gets its wonderful stickiness from a density of saturated fats so great it would make a meal of ordinary butter slabs seem as low fat as nibbling lettuce leaves. A single medium-size bucket of popcorn buttered with this coconut oil will, according to nutrition analysts, "contain more saturated fats than a breakfast of bacon and two eggs, a steak dinner with side dishes, and a Big Mac with a large order of fries—combined."

Sharing slobbery big handfuls of popcorn from the same bucket leads to a curious reincarnation for the bacteria we pick up—left on the nabbed-for kernels from the friend's saliva—in the process. They are destroyed when we swallow them, but that's not the final end. The raw chemicals that compose the bacterium are largely salvaged inside the body; our liver enzymes easily stripping away the few bits that are truly dangerous to keep away. The remainder flow back, pumped to the portions of our throat or elsewhere where yet another generation of these gigantic monsters are being grown. When these new bacterium get handed across during next Saturday's adventure installment—the two friends perhaps bemoaning the lack of interesting things going on in the movie—they briefly metabolize in their new home, only to be swallowed back by the body they started from, ready to be regenerated, and the great cycle starts again.

What there will be on the screen, almost certainly, is an explosion. Or several explosions. About 60 percent of the most popular films of the past ten years have featured this apparently irresistible destruction of solid objects. There's also likely to be a fight (in 62 percent of PG movies) and then, lurking somewhere around on the bright screen, there's going to be a guy named Jack. This is the most common name in movies from the past decade, by far. It's not always reserved for the hero—sometimes Jack is discreet, and settles for a bit part, letting the main character have a different name—but it's a rare film where a Jack of some sort isn't there at all. With a stunning lack of linguistic variation, the next most common name is John.

What Jack, John, and the other huge projections on the screen will do when they're not fighting or watching explosions, is say simple things. The phrase "Let's get out of here!" is exceptionally common, according to one determined English researcher who viewed over 100 of the main U.S. feature films of the last half century. It was uttered, exclaimed, whispered, or otherwise conveyed in 84 percent of them. The characters in movies also swear, though this is carefully adjusted for the intended classification level on release. Many thrillers are filmed with key swearing scenes done as written for the R movie release, and then, while everyone is still on the set and the camera is in place, those scenes are refilmed with more innocuous words. The screen shadows kill people too—there will be at least one murder in 40 percent of PG and G films—but not as frequently as they swear or fight.

There is one slightly less obvious message displayed on all screens, and you can see it in the upper-right corner, at about forty-minute intervals. The message is there to tell the projectionist that a reel change is coming up. A squiggle appears brightly on the screen when there are about twelve feet of film left, and then, seven seconds later, when the projectionist is down to two feet of film, it pops up on the screen again. Directors often draw out the end of a scene till after this changeover point, so no important dialogue will be lost if the projectionist blows it. Once you've had the squiggle pointed out it's hard to miss it: the thing is big enough for the projectionist to see clearly from his distant booth. To find it just wait till there's the giveaway slight jump on the screen where the reels were changed. If the boys know how long into the movie that was, they just have to wait the same amount of time again, then look up from their life-cycling popcorn, and there it'll be.

The girl would love to escape reality in a movie, but instead is alone in a mall bathroom, dispiritedly looking at herself in a mirror, dabbing at her face with tissue. She knows you're supposed to feel better after a cry, but she doesn't, really, even though a great number of the stress chemicals that had been surging in her body were pumped out with the tears. She's also automatically helped by the tear glands opening into the bottom of her eyelids. They've been pulling in the ordinary low levels of antibodies from her bloodstream, especially the all-important immunoglobulin-A variety, which is what the baby received in its protective breast milk. The tear glands concentrate those antibodies into denser and denser agglomerations, and when the antibodies are ready the tear glands start pumping them over her eye, too. But this is no great comfort either; not after the desolation of being unpopular; so clearly an outsider even among those she considered her friends.

She tilts her head back now, the dropper from the bottle of eye lotion she bought held up at the ready. It was easy enough for her to find this bottle at the pharmacy on the way here, for it's a good product to sell. What's in the dropper is usually a generous 50 to 70 microliter pool, and most of that, once carefully squeezed down, will immediately pour down to the bottom lid, either to overflow down the face, or to be sniffed uselessly through the drainage holes leading from the nose. Our eyes can only keep about 10 microliters of extra fluid on them at once, as the manufacturers of these bottles well know. The poured-away extra is simply a quick way for them to get the bottle emptied. The little bit that sticks on the retina—there's a little polyvinyl alcohol, the stuff on the back of stamps, to help—is useful though, for it soaks down into the living eye, binds to the cells of the swollen blood vessels there, and makes them shrink.

A final tissue dab from the girl, and it's time to go. In the remaining hour till she's going to meet her parents she'll do some shopping, or maybe—for what is it ever going to matter how she looks?—just get some chocolate, yeah, lots of it. She opens the door.

And immediately bumps into a boy, the sensitive one with the (really quite slight) case of acne from the earlier group.

He's intently saying something about the other kids, but she can barely hear him: there's suddenly an incredible roaring in her ears. He must have been waiting for her, on purpose, and that means he has to be interested in her, and that means they might end up going together, and then everyone will expect that she's really involved with him, and he'll think that too, but that's terrifying because she's never slept with a guy, or really done anything with a guy. She's never—the one absolute secret, which she has guarded from her folks, and especially from everyone at school, she's never even *kissed* a guy.

Which is only understandable, being the eldest child in a family. Eldest children are almost always more restrained than others. In a study where teachers were asked to point out which were the most physically adventurous kids in a classroom, the ones they selected usually turned out to be youngest siblings; when they were asked to point out the most cautious ones, they were kids who were the oldest in their families. Eldest kids are the ones most likely to obediently take on their parents' professions, be it medicine or accountancy or living on a farm or even, as statistics back up, being an Episcopalian minister. In science, although there have been notable exceptions, first borns are usually the ones most likely to prefer long-established, traditional theories. Most of the resistance of established scientists to new theories such as special relativity or continental drift came from those who'd been the eldest in their families, too.

When it comes to sex, the same cautiousness holds. The average age for first having sex in America is now about seventeen, but it's not because everyone starts at that point. In a family of two kids, the average will likely derive from the older one starting at seventeen and a half, and the younger at sixteen and a half. If there are three kids, the youngest one will end up starting even earlier, and if the spacing between the kids is relatively wide, the difference is wider, too. Despite the daughter's pretense of social independence and utter sexual insouciance, her parents can take heart that she's unlikely, without strongly veering from the statistical average for eldest daughters in families of three kids, to start having sex before age eighteen and a half. It's the sweet-as-sugar youngest kids in big families the parents have to watch out for later. Their average age for first having sex is barely past fifteen in America now, and dropping. The combined result is that about 1 million U.S. teenagers will become pregnant this year. This sounds like a lot, but it's been higher. The peak year for teenage pregnancy was actually 1957 . . . but that was simply because the average marriage age was but nineteen.

Despite her hesitancy the girl starts walking with the boy; at first still cheek-burningly self-conscious, still embarrassed at what happened before. But gradually, as they window shop and chat, and he's gentle in all his words, she forgets that she's supposed to show off or keep him at a distance, and then—lost in their shared companionship—she just really likes it. They laugh at a hurrying family group and then at a funny window display, and suddenly it's even better, for with laughter cathecolamines start spurting inside them. These are a category of chemicals that start cascading from the brain in ordinary laughter, and act on cells in the circuits concerned with alertness and attention. The result is that the shared walkers, without quite knowing why, will suddenly feel even more closely attuned to each other.

They'll also feel happier, as several trillion of the morphinelike endorphin molecules start spraying loose within their brains. Endorphin's effect is so powerful that if the girl had been subconsciously digging her fingernails into tightly closed hands, or if he bumps into a bench edge while demonstrating a mock dance step, neither will feel any pain signal, as the endorphins will be active in their spinal cords to block it.

It's hard to tell how much time has passed—there are rarely any clocks in shopping malls, intentionally to keep shoppers isolated—but what does it matter? The couple are together on these timeless walkways. It's only natural for him to put his arm around her, and for her to lean against him, and now everything feels very very different, and the world of parents and brothers and breakfast squabbles is suddenly unimportant; immensely far away. Somehow they end up taking the turnings toward quieter corridors, no longer scared of what might happen when it's isolated enough to stop.

Walking this closely, the invisible world of sexual vapors lifting off their bodies, far stronger than their parents' immune clouds, is helping them along. The boy is spraying out the fatty chemical called androstenol from his surface blood vessels and especially from his armpits. This isn't the ammonia vapor you get from unwashed clothes. Rather, when it's concentrated enough it's usually described as having an enticing sandalwoodlike flavor. It's floating over to the chest-nestling girl at levels too low for her to consciously notice, but this doesn't mean it's having no effect. When college girls look through pictures of male students, if the level of air-filling androstenol goes up, the amount of interest they say they feel toward the man in that picture goes up, too.

The sounds of other walkers are increasingly dim, and so finally, beside an unused stock entrance door, off on the side of the upper level, they do stop and face each other, trying not to be to awkward as they entwine arms, and then, eyes closed, they tilt their faces close. The mucosal folds designed to keep food from spilling loose pucker outward, and then—after only the slightest of final life-spinning hesitations—their lips finally touch. Many cultures have resisted this strange fashion. In Japan it was abhorrent until 1945, but America's military occupation authorities encouraged it—even forcing screenwriters to include it in their scripts—on the grounds that anything encouraging young people to make their own decisions and not follow their elders (America's recent foes) was good. Nerve receptors are compressed and tugged in our couple, and data streams, many more than usual, travel at 80 mph along pathways skirting the outside of the cheek, then up the few inches into the sensorimotor sections of the brain. Bacterial transfer is still very slight at this point, for light contact only touches the outer surface of the lips, which are bacteriologically quite clean. Lips are a frontier zone between the skin's microbial lifeforms and the quite different ones inside the mouth, so although a few mutated forms manage to colonize the lips, much of the surface area is empty, as each arriving species battles the others to oblivion.

A little more pressure though, to really contact the beloved, and now the

The couple-friendly nuzzling of a kiss, viewed via heat-sensitive film.
The necks and foreheads glow hottest white; shoulders and nose are cooler,
shown in dark blue.

mucosal folds start squashing open. A low-atmospheric suction tunnel is created, linking the pair, and the first saliva streams begin cross-sloshing, whipped up by the sudden internal gale. Outer levels of bacteria are ripped loose from the teeth they're clutching to, the cementlike ligand molecules are incapable of holding them in place against this wind tunnel–like onslaught. The stored sebum from the normally subterranean follicles at the corner of the pressing lips are ripped upward, and with it, tumbling in, go great spurting jets of flying acne bacteria.

It could be enough to give anyone pause, but who can resist something that feels this trusting; this close? Desire helped by the fact that now there are yet more powerful sex chemicals hovering in the air between our delighted couple. The androstenol on the boy's body has been changing, aided by the raised temperature that kissing produces. Some of it has transformed into the even more potent androstenone molecules, and although men can barely smell it on themselves, and strongly dislike it if they ever do get a concentrated whiff from other men (they'll avoid a chair in a waiting room that has been coated with it) women in contrast like it, very much. They'll head *toward* a waiting room chair that's been dabbed with it, especially if they're given this test at the halfway point in their monthly cycle, when fertility is at its peak—and, as we saw, when the girl's blood-testosterone levels and libido would be highest, too. (There's a little androstenone in musk aftershaves—put in using a chemical process developed by none other than Wallace Carothers, the nylons inventor, which perhaps has a certain symmetry.)

Human beings aren't simple chemical machines, but at moments like this, drunk with the mix of crackling brain circuitry and androstenol clouds, we come pretty close. A practice gleaned from a certain country whose capital is Paris might even be engaged in. Mouths open slightly wider, and at this point, anyone of a squeamish disposition who has not already averted his eyes should probably do so. Tongues are extraordinarily life-dense objects, for on their roughened surfaces exists a safe refuge for innumerable microbial colonies that would otherwise be removed by ordinary chewing or saliva flow. The tongue is especially dense in the primitive anaerobic species, that normally live cowering deep underneath the surface layers, and that evolved deep in swamps and other areas where they could hope to exist away from their great nemesis—free-floating oxygen. Now, with the great lingual exploration beginning, they are scraped loose and wildly flung up. Yeasts and spirochetes and other mobile bacterium are shaken loose, too.

But readers who averted their eyes will have missed an unexpected sharing, something quite touching at this point. Until now, each of the teens has shared their separate family's general defenses against the external world's biological entities. But now, fresh biological linkings, paralleling their tentative emotional contacts will also, for the first time, intentionally be underway. Most of the anaerobic bacteria survive the impromptu aerial voyage, and hunker in their new home on the strange new tongue just fine. The giant *T. rex* of the mouth cavity, the great marauding *E. gingivalis* beast, is also likely to be making its way across, if the boy has brought it to the mall; either swept up in the general ruckus or migrating by its own power over the now extended tonguely bridge. This arrival in the girl's mouth is not any great problem, for it will simply continue its actually quite useful task of clearing up the micromeat of tooth-threatening bacteria that might already be inside her, resident on her gums. And as an example of the reciprocity this kindness entails, even as the nerve tinglings and brain contact continues, the girl is nearly certain to be loading her guy now with live colonies of *veillonellae* bacteria, which she, ultraconscious of hygiene, is likely to have nurtured at greater densities than he. The *veillonellae* take up residence in him, there amid the tumbling-down anaerobes, and diligently get to work soaking up the otherwise dangerous lactic acid, that his own, male-hygiene levels of mouth bacteria are likely to have let loose inside him.

Where there has been too much transfer, this is automatically cleared up, even as the new pair continue their delighted, boundary-crossing bliss. Saliva production goes up with the excitement of being kissed, and that means more lysozyme chemicals to break open the walls of excess bacteria, and more mucin proteins now coating the girl's mouth to stick other excess ones into harmless, dot-sized lumps. Each swallow will drain millions to the stomach for quick hydrochloric destruction, and even each gasp of pleasure will spray extra oxygen inside to neutralize any extra anaerobes still doing their free-fall gymnastics, turning them into crash-landing dying fragments. And where levels are still not right, and too many of the beloved's microbes have still made it into position? The extra saliva of this embrace will also be flowing with rich supplies of extra bicarb, and that, floating over to the attacked positions, quenches the acid the microbes produce. When the bicarb hits the acid, gusts of carbon dioxide bubble upward, tumbling to the air ducts and soon on to the world outside, yet still holding the trace of this isolated, enraptured pair.

epilogue: late night at home

Back at home, the girl contentedly perched on the soft chair across from the TV, her mouth continuing to float out carbon dioxide bubbles as she clears up any possible excesses from the afternoon's intimate transfer. The metal within her dangling bracelet acts as an inadvertent antenna, soaking up the high-speed radio waves that whirl into this room from distant stations, spinning the waves in roaring fast arcs around her wrist. As the bracelet lacks any amplifiers, all the fast-talking voices and insistent love music will quickly fade out of existence—which doesn't matter to her, as the girl is replete with wondrous reality itself. She just goes on dreamily musing, rolling over in her mind the delicious moment when he phones tomorrow and she lets her mom answer and someone calls up the stairs, in a hurried whisper, that it's a *guy* for her; she can answer, oh yes, she knows, it must be her *boyfriend.* Macrophages in her lungs are still trying to tug bagfuls of the smoke-burned polycyclic out of harm's way, and each breath out releases a few thousand of the cigarette-produced cotinine molecules, which tumble down her arm, then bounce in easy slow motion over her unheard bracelet to float on into the room. Higher up, on her face, demodex follicle mites are cautiously emerging; the excess makeup that locked them in now washed off.

The ten-year-old brother has no idea why his sister is being so nice— she even let him grab the best window seat on the way home—but he's not going to complain. She's letting him take the couch, for one, where it's great being the man of the house and they're watching his choice of video. She's made him a peanut butter and jelly sandwich, which she hasn't done in maybe a million years, and she's even brought him a second glass of milk, too. Sugar-hunting bacteria have been accumulating on his teeth, from the day's barrage of maple syrup, soft drinks, candy bars, Cracker Jacks, jelly beans, peanut-butter buttons, milk chocolates, caramels, popcorn, and that chewing gum. An evening's glass of milk is an excellent final defense against them all. One group of proteins especially abundant in milk have been keeping the bacteria from finding places to get properly attached; another chemical has been capturing some of the torn-off enamel that earlier bacteria had pulled off, and collecting it to be led back into place. He's feeling better inside as well, as his liver's detoxification systems have finally disposed of the burger's unwanted hormones, leading them safely into his bladder to join the aluminum, food additives, and even fragments of the vast eye-landing rubble from the air that had been blinked through the connecting tunnel to his nose, dissolved in his stomach, and now in part ended floating here, too.

A scary moment in the video, and the girl reaches to take a sip of his milk. The marauding *E.gingivalis* micromonster that her brother has left on the glass comes alive, dimly detecting this arriving warm flesh, and slowly, laboriously, stretching out one roiling amoebal arm to try to find it. But the girl has her own marauding specimens to leave on the glass edge, too. As she puts down the glass, the two beasts will still be a great distance apart, but these individuals are almost impossible to kill, and later this night, if the glass is left out unwashed, they will survive, slowly stumbling forward. The meeting will end either in a final slow-motion attack between these ancient ones, or, if the encounter is of a different sort, in the production of eight doubly filled cysts, which a few hours later will produce a squelching *Alien*-like expulsion, and sixteen living baby amoebae, waiting for whoever comes down earliest tomorrow, to rewind the video, and—could they?—pick up the glass, for a final lip-touching sip.

On the floor around the TV-fascinated brother and sister, the carpet is quietly digesting the day's accumulation of tumbled yeasts, spores, dandruff particles, and spilled food crumbs, as well as the final landing skin flakes, along with the dense cloud of pet saliva proteins. The dog which generated them is lying fast asleep on this active floor, nostrils only slightly blocked by the residual air freshener chemicals drifting out from the opened stick on the shelf, and absorbing the occasional radioactive radon gas oozing up from the floorboards. Ozone gusts from the upper atmosphere are still slamming against the house, but the DNA-linked ants which had been walking on the outside bricks are no longer there to be assaulted, but rather are snuggled deep in their subsurface nighttime nests. Far beyond, the expanding bubble of radio and TV waves from earth have continued moving outward, with this day's earliest broadcast offering (distant aliens are safe from cable) now itself about 8 billion miles away and going strong.

In the empty kitchen, the Formica table is slowly evaporating upward, while a few fragments of the mornings N_2O gas remain hovering shyly above the stove, not yet started on their long earth-circling voyage. Most of the flying fungi from the under-sink cupboard have landed, oblivious to the slow, nighttime ethylene gases that the apples keep on floating in painful communication to each other, as well as to the constant sizzle of radon gas, pushing up from the basement below. Over it all, the sturdy smoke detector is back to its disappointing, consistently null readings, each radioactive americium-241 spurt it beeps out registering only the faintest smoke rubble laggards.

One level up, the parents are in the bedroom, sitting at the wife's desk where a photo album is spread open to glue in—so that this time they do not get caught in a backlog—the newly processed pictures. They're feeling tired, which is understandable, as they're inhabited by brains that are now at the end of a day full of sorting input data, blocking poisons, puffing out carbon dioxide, suffering bits fading out through repair failures, and living on sugar and oxygen, all the while sloshing in wet fluid within the cranium. Percloroethylene continues gusting out from the clothes closet, but much has already been suctioned away by the micropores on the back of the now-tired geranium. The geranium is slowing its respiration rate at this late hour, after the effort of processing those vapors and then injecting them into the circulation pathway that will lead, as the parents are finally asleep later tonight, all the way down to the plant's roots. Numerous pillow mites have been returned to the bed after their venture out today, with a few live specimens from other

The inner mechanisms of a wristwatch signal the end of another day.

homes added, trying to take up new homes there, safe from the lurking *Cheyletus* predators.

Carbon dioxide is wafting in from the brain-churning children downstairs, sitting before the video, and reactive oxygen in the air is already starting one of the processes that will ultimately send the photographs' life-bursting images fading gradually away. The genetic links the family's creating will also gradually fade. By the mid-2500s, as the last of the CFCs released from this morning's kitchen are disappearing in the upper atmosphere, they and their children will only exist as tiny DNA fragments, wedged here and there in the widely separated cells of about 2^{20} or over 1 million great-great-great-great-great-great-great-great-great-great-great-great-great-great-great-great-great-great-grandchildren (less depending on inbreeding, mutation, or genetic drift).

Suddenly there's a cry, a shout almost, but different from what's been heard in this house before. Any maudlin thoughts are abandoned, as the parents hurry along the electron-scuffed trailmarks of the upstairs hall. In the final room along that hall, the baby's bedroom, their ten month old—this youngest creator of future generations—hasn't been asleep at all. Instead, he's pulled himself up, holding on to the edge of the crib for balance, concentrating intently. There had been something about that shiny helium balloon, tethered to the changing table on the other side of the room, that had made him struggle to hook his brain cells into an even more complex configuration than ever before.

And then, in the half-light, he gets it again.

The baby's excited voice shouts out once more, and the parents realize—and what greater joy in life is there?—that they're now hearing its first blurted word. They hold hands, delighted, here in the faint light of the nightlight, and they step forward, one more shared journey, to hug their stupendous child.

And are even quite polite, at first, as they discuss whose name *exactly* it was their son had just said.

photo credits

Dr. Ray Clark/Mervyn Goff/MMPA AIIP ARPS/Science Photo Library, frontispiece, 22, 202

David Parker/Science Photo Library, 29

Science Photo Library, 34, 45, 167, 210

Prof. P. M. Motta, G. Macchiarelli, S. A. Nottola/Science Photo Library, 41 (top)

K. H. Kjeldsen/Science Photo Library, 41 (bottom)

Andrew Syred/Science Photo Library, 42, 69, 101, 172

David Scharf/Science Photo Library, 48, 57, 113, 151

Dr. Jeremy Burgess/Science Photo Library, 53 (top), 73, 163, 181 (both)

Dr. Tony Brain/Science Photo Library, 53 (bottom)

Clive Kocher/Science Photo Library, 60

Space Telescope Science Institute/NASA/Science Photo Library, 65

Manfred Kage/Science Photo Library, 76, 117

Natural History Museum, London, 79

CNRI/Science Photo Library, 83, 157, 187

Biophoto Associates/Science Photo Library, 86

D. Philips/Science Photo Library, 90

Princess Margaret Rose Orthopaedic Hospital/Science Photo Library, 96

Dr. Gary Settles/Science Photo Library, 108

Dr. David Gorham and Dr. Ian Hutchings/Science Photo Library, 119

Prof. P. Motta/Department of Anatomy, University *La Sapienza,* Rome/Science Photo Library, 125, 144

Eye of Science/Science Photo Library, 130

Richard Wehr/Custom Medical Stock Photo/Science Photo Library, 141

Department of Clinical Radiology, Salisbury District Hospital/Science Photo Library, 146, 194 (both)

M. Marshall/Custom Medical Stock Photo/Science Photo Library, 147

AGFA/Science Photo Library, 177

index

Page numbers in italics indicate illustrations.

acetylation, 36–37

acetylcholine, 169

acne, 27, 126–27, 203

additives:

    food, *see* food, additives in

    wood, 49

adolescents, 39–40, 42–43, 176,
      178–79

    acne in, 27, 126–27

    dieting by, 165–66, 168–69

    homework time of, 122

    Muzak and, 148

    relationships between sexes of,
      179–80, 182, 199–201, 203–5

    smoking by, 182–83

    weight of, 124

adrenaline, 138

adrenocorticotropic hormone (ACTH),
    43

advertising, television, 111

aging, 185–88

AIDS virus, 149

air pollution, 56, 140

    plant mechanisms for absorbing,
      71–72

air travel, 139

alkaloids, 98–99

allergies, 91, 112

alpha waves, 40

aluminum, 36

Alzheimer's disease, 137

Ambrose, St., 83

American Muzak Corporation, 148

amino acids, 95, 184

ammonia, 50–51

anaerobic bacteria, 203–4

androstenol, 201, 203

anger, 135–36

antibiotics, 150

antioxidants, 161, 169

arguments, 173–75

Aristotle, 40, 185

asbestos, 115

Ashdown, Paddy, 153–54

aspirin, 126
auto-brewery syndrome, 193, 195

babies:
    blink mechanism in, 55
    brains of, 20, 32, 115, 136, 156, 184,
        189
    breast-fed, 155–56
    breathing of dust by, 114
    caffeine and, 175–76
    comforting, 128
    DNA repair systems of, 49
    docking position for, 30
    exploring, 88, 89, 91–93, 105
    face recognition structures of, 136
    first words of, 211
    food for, 19–21, 23, 47
    lead absorbed by, 115
    parent-controlling behavior of, 39
    pillow mites and, 130–31
    premature, lungs of, 110–11
    sweat glands of, 142
    talking to, 184–85
Babylonians, ancient, 62
Bach, Johann Sebastian, 40
Backus, Peter, 74
bacteria
    acne, 126–27, 203
    airborne, 114–15
    defenses against, 74–75
    demodex mites and, 27, 28
    in food, 82, 196–97
    kissing and, 201, 203–4
    in public spaces, 149–50
    in soil, 66
    on stamps, 100
    on teeth, 75–76, 145–46, 208
    on towels, 127

baldness, 185
balloons, 184
bathing, 119–24
beef, 158–59
beta-carotene, 161
biodegradable plastic, *167*
biorhythms, 148
birth order, 199–200
bladder, 36, 88, 189–90, *194*, 195–96
blink mechanism, 55, 90, 149, 191
body fat, 124
    cells, *125*
    hunger and, 156–58
    in middle age, 186
bone, 139, *144*
    growth of, 178
brain, 209
    aging and, 188–89
    of babies, 20, 32, 115, 136, 156, 184,
        189
    of children, 85–86
    clearance cells in, 137
    during concentration, 69
    dog petting as stimulus to, 51–52
    gender differences in, 33
    of insects, 98–99
    language processing in, 83
    memory function in, 39
    nerve signals received by, *79*
    pleasure receptors of, 64
    power rating of, 30
    protection from toxins of, 36
    scans of, *83*, 85, 169
    smiling and, 154
    sound recognition in, 120
    stress and, 43
    tryptophan in, 95
    visual centers of, 38, 85

Brantôme, Seigneur of, 174
bread, 43–44, 50
breast-feeding, 155–56, 175
bubbles, *29*, 122–23
bunchu, 168
butter, 47, 54–55

caffeine, 99, 101, 175
calcium, 168
carbon dioxide, 110, 169, 210
carbonated soft drinks, 160
careers, 186
    birth order and, 199
carotene, 161, 165
Carothers, Wallace Hume, 93, 203
carpets, dust in, 111–16, 208–9
cars:
    buying, 77–78
    exhaust fumes from, 71, 140
cathecolamines, 200
cats, 52
    blink mechanism in, 55
    dried saliva of, 112
    peripheral vision of, 62
cells
    bone, 139, 168
    brain, 137, 188–89
    fat, *125*, 156–58, 166
    life span of, 149
    memory, 149–50
    nerve, *60*
Chanel, Coco, 66
chemicals:
    dry-cleaning, 103–4
    in foods, 36–37
    in plastics, 43
    in water, 52
chewing, 62

chewing gum, 145
Cheyletus, 129, 132, 210
chickens, 46–47, 162, 164
chlorine, 121
chloroflurocarbon (CFC), 32, 56, 58,
    210
chocolate, 193
cholecystokinin, 184
cholesterol, 71, 101, 184
chromosomes, *86*
cigarettes, *see* smoking
Clinton, Bill, 153
clothing, 138–40
    shopping for, 189, 192–93
coffee, 99, 101, 175
cold viruses, 154–55
collagen, 123
colors
    of clothing, 139–40
    of cosmetics, 121–22
    perception of, 89, 90
    shopping behavior and, 191
communications systems, 135, 153
compact disc, *53*
computer chip, *73*
computers, 85–86, 97–98, 102, 111
contraceptive pills, 25
conversation, 38, 153
cooking, 54–56
corpus callosum, 33
cosmetics, 176, 178
    colors of, 121–22
    shopping for, 192
cosmic rays, 87–89
cotinine, 92
counting rhymes, 95
crying, 183, 198
cystic fibrosis, 75

Danish pastry, 73–74, 80–82

Darwin, Charles, 97

demodex mites, 27–28, 30, 123, 126, 128, 129, 207

Dermatophagoides pteronysinus, 129, 132

detergents, 87
    bubbles, *29*

detoxification, 36, 49, 145

dextrin, 44

diamonds, 96

diatoms, *172*

Diderot, Denis, 139

dieting, 165–69
    smoking and, 183

digestion, 63
    dieting and, 165–66
    of vitamin pills, 37

dioxin, 54, 166

divorce, 33, 138, 173

DNA, 23, 180, 211
    repair systems, 49, 65, 88, 89, 161

dogs, 51, 63–64, 209
    breeds of, 67
    caffeine and, 175
    Entamoeba gingivalis in, 76
    hearing in, 61, 71

Donne, John, 185

doorbell, 59, 61
    sodium in battery of, 60

doorknob mechanism, 64

dreams, 26
    waking, television watching and, 107

dry cleaning, chemicals used in, 103–4

dust, 106, 111–14, *113*

eggs, *41, 42,* 44, 46–47

Eisenhower, Dwight, 102

electric power, generation of, 31

electromagnetic waves, 109, 120

electronic pacemaker, *145*

elevators, 143–44, 147–49, 152

Elizabeth I, Queen, 178

endorphins, 64, 200

Entamoeba gingivalis, 76–77, 204, 208

Erasmus, Desiderius, 39

esophagus, 63

estrogens, 40, 43, 182

eyeglasses, 182

eyes
    bags under, 185–86
    drops in, 198–99
    drying of, 97–98
    horizontal scan of, *83*
    reflexive widening of pupils, 20, 84–85
    repair mechanisms in, 88
    rotation toward sound source of, 61
    sunlight and, 64, 139
    during television watching, 107–8
    see also vision

face recognition, innate, 136

Factory Outlet Marketing Association, 189

families:
    configuration of, 33–34
    isolation of, 62–63
    names of, 69–70
    personal space of, 136

fat cells, *125,* 156–58, 166

fear, 136

fibrinogen, 71

fingernails, 121

fingerprints, 88–89
fingers, 97
    joints of, 171–72
fish, 162
Fletcherizing, 62
flossing, 76
fluorescent lights, 32, 143
food, 145
    additives in, 36–37, 47, 80–81,
        208
    baby, 19–21, 23, 47
    brain activity and, 169
    children's choices of, 82
    cooking, 54–56
    digestion of, 63
    microwaving of, 73–74
    packaging of, *167*
    refrigeration of, 30–32
    shopping for, 190
    vitamins in, 37–38
    see also specific foods
forks, 161
formaldehyde, 36, 49, 100
fresh air, 56
fungi, *48*, 50
Funk, Casimir, 37–38

galactomannan, 80
gas chromatography, 103
gender differences:
    in approach, 68
    in bladder size, 190
    in body perception, 166–67
    in brain, 33
    in finger length, 97
    in heart disease, 107, 124
    in looking in mirrors, 143
    in medical knowledge, 188
    in personal space, 135
    in purchasing decisions, 77–78
    in reflexive widening of pupils, 20,
        84–85
    in school achievement, 122
    in sensory perception, 61, 81–82
    in shopping behavior, 189
    in smile perception, 154
    in sweating, 142
    in telephone answering, 102
    in treatment of babies, 105
genes:
    for acetylation, 37
    cholesterol and, 71
    hay fever and, 91
    macrophages and, 114
    of siblings, 26
glue, postage stamp, 99–100
glutathione, 140
gold, 96
goosebumps, 142
grapefruit, 101–2
grass, 72
Greeks, ancient, 44–45
Greeley, Andrew, 175
guitar string, *101*

Haber, Fritz, 51
hair:
    mites on, 27, 28
    red, 24
hamburgers, 158–60
    tapeworms in, *157*
handedness, 138, 190
handshake, 153, 154
handwashing, 190
Harrison, James, 31
hay fever, 91, 112

hearing, 61–62
    in dogs, 71
heart disease, 71, 107, 124
heat, body, 140
hemoglobin, 50
Henry, Edward, 89
hexanal, 49
hippocampus, 85
Hippocrates, 44
Holmes, Larry, 78
homework, 122
homunculus, sensory, 79
hormones:
    in adolescence, 178, 180, 182
    aging and, 186
    female, 43, 54
    stress, see stress hormones
    weight and, 124
house addresses, 67–68
houseplants:
    hemoglobin-related chemical in, 50
    oxygen released by, 24
    toxins absorbed by, 49, 209
hunger, 156–58

Iacocca, Lee, 169
ice, 31
ice cream, 184
immigrants, 138
immune system, 149–50, 186
    cold viruses and, 155
    mate selection and, 24–25
    pillow mites and, 130
    of siblings, 26
    stress and, 43
incandescent bulbs, 32
incomes, disparity of, 138
insects, 66, 98–99
insulin, 95, 156

interracial marriage, 179
ionone, 142
IQ:
    of breast-fed versus bottle-fed babies,
        155
    eyeglasses and beliefs about, 182
    in mate selection, 25
    of siblings, 26
isoclines, abdominal, 186

jewelry, 176, 207
joints, 171–72
junk mail, 77–80

kahweol, 101–2
ketchup, 159–60
Khmer Rouge, 44
kidneys, 36
Kikuyu, 40
kissing, 199, 201, *202*, 203–5
knuckles, cracking, 171–73

language:
    brain areas for processing, 83
    infants and, 20
laughter, 183, 200
lead, 52, 115
Lee, Robert E., 153
Leonardo, 161
lettuce, 161–62
lightbulbs, 32
lighting:
    emergency, in movie theaters,
        196
    voice level and, 179
    see also sunlight
limbic system, 154, 183
lipstick, 178
literacy, 84

liver, 188
    detoxification in, 36, 49, 145, 208
Louis XIV, King of France, 63
lungs, 75
    during television watching,
        110–11
    house-cleaning cells in, 114–15
    smoking and, 183
lysozyme chemicals, 149, 204

macrophages, 114–15, 183, 207
magnetism, 44–45, 88
makeup, see cosmetics
malaria, 118
Mao Zedong, 63, 76–77
marriage, 24–25
    cultural differences in, 97
    disagreements in, 173–75
    duration of, 34
    frequency of sex in, 174, 175
    interracial, 179
mate selection, 24–25, 180
meat, 156
    dogs as, 52
    in baby food, 21, 23
Medawar, Peter, 176
medical knowledge, 188
melanin, 50
melatonin, 64
memory, 38–39, 85, 169
Memory B's, 149–50, 154, 188
men, differences between women and,
        see gender differences
menarche, 40
meningitis bacterium, 149
menstrual cycles, synchronized,
        179
microwaves, 73–74, 80
middle age, 185–88

milk, 95–96
    ability to digest, 160
    mouth bacteria and, 208
mirrors, 143
mites:
    pillow, 129–32, 209–10
    see also demodex
mosquitoes, 109–10, 116–18, 117
moth effect, 158
mouth:
    bacteria in, 75–76
    life span of cells in, 188
mouth-feel, 80–81
movies, 195–98
Mughal empire, 31
music, 120
    eating rate and, 164
    elevator, 148
    in stores, 191
Mussolini, Benito, 153

names, 69–70
    in movies, 197
    popularity and, 182
Nash, Ogden, 43
nerve cells, sodium in, 60
neurotransmitters, 26, 96, 98
nicotine, 92, 182–83
Nielsen Media Research, 108–9
nitrogen molecules, 89
nutrition:
    and onset of puberty, 40
    see also food
nylon, 92–93

omentum, 186
orange juice, 35–36, 54
osteoblasts, 139, 168
ovaries, 41

overweight people, 156–58, 164

ozone, 91, 105, 209

pancreas, 95

pencil, *69*

penicillin, 126

Pepsi company, 166

percloroethylene, 103–4, 209

peristalsis, 63

personal space, 135–36

   in elevators, 147

pets, 51–52

   see also cats; dogs

phenylethylamine, 193

phonograph record, 53

photons, 64, 139

phytic acid, 168

pillow mites, 128–32, 209–10

pineal gland, 64

pitorosporum ovale, 28

pizza, 169

placebo effect, 38

plants:

   defense systems of, 66–67, 106

   food reserves of, 161

   toxins absorbed by, 71–72

   see also houseplants

pollen, *90*, 90–91, 112–13

polycyclic hydrocarbons, 183, 207

polymers, 21

polyphosphates, 164

popcorn, 195–98

postage stamps, 99–100

potato, *163*

prayer, 175

pregnancy, teenage, 200

propriabacterium acnes, 27

puberty, 40, 124, 127

Quayle, Daniel, 154

racism, 138

   reduction in, 179

radio waves, 120, 207, 209

radon gas, 49, 92

reading, 83–85

   computer screens, 97–98

reflexes, 61

   widening of pupils, 20, 84–85

refrigeration, 30–32

salad, 161–62, 164–65

saliva, 145–46, 204

scales, bathroom, 124–25

sebum, 127

sensory homunculus, *79*

serotonin, 95

sex:

   adolescents and, 180, 182, 199–201

   frequency in marriage of, 174, 175

shopping behavior, 189–93

sialin, 146

sidedness traits, 82

silicon breast implants, *146*

singing, 120

skin:

   blood flow to, fear and, 136

   detergent residues and, 87

   dust emanating from, 106

   exposure to sunlight of, 139

   pressure receptors in, 116

   wrinkling of, during bathing, 122

sky, 90

sleep cycle, 26, 43

smell, sense of:

   of babies, 49

   of dogs, 51

   food and, 81

gender differences in, 82

and mate selection, 24–25

smile, fixed, 154, 182

smoke detectors, 55–56

smoking, 92, 152, 182–83, 207

sneezing, 91, 154

soap, 123

sodium, *60*

soft drinks, 160

sound, 59, 61–62

speech patterns, 38

sperm:

deformed, 104

production of, decreased, 43, 52, 54

spinach, 164–65

spinal discs, compression of, 186

spirochaetes, 76

stars, exploding, *65, 96*

stomach, 63, 165

bacteria in, 75

smoking and, 183

strawberry, *57*

street names, 67–68

Streisand, Barbra, 153

stress hormones, 43, 135, *176*

in adolescence, 178, 180

during arguments, 174

crying and, 198

submissive posture, 179

sugar, 160

babies and, 184

bacteria and, 208

in chocolates, 193

sunlight, 64–65, 87

eyes and, 139

plants and, 161

swallowing, 62, 63

sweating, 88, 140, *141*, 142

Swift, Jonathan, 46

tapeworm, *157*

taste, sense of, 61, 81–82

Taylor, Elizabeth, 153

tear glands, 198

teeth, 62, *76*

bacteria on, 75–76, 145–46, 208

telephone use, 82, 102–3, 128

television watching, 107–11, *108*

testicles, protection from toxins of, 26

testosterone, 33, 180, 186

thermogram, *22*

three-dimensional orienting, 33

time management, 98

time measurement, 62

toast, 44

tobacco, *see* smoking

Tocqueville, Alexis de, 174

toenails, 121

tongue, 62, 203–4

toothbrushes, 127–28

towels, bacteria on, 127

trees, 66–67, 71–72

trichosiderin, 24

Twain, Mark, 52

ultraviolet rays, 64, 87, 139

urinary system, *see* bladder

Valle, Pietro della, 99

vasopressin, 186

vegetables, 161–62, 164–65

veillonellae bacteria, 204

Velcro, *181*

viruses, 154–55

vision, 38, 152

aging and, 186–89, *187*

concentration of, 69

of infants, 32

vision (*cont.*)
   peripheral, 61
   vegetables and, 165
vitamins, 37, 65–66, 139, *151*
vocabulary, 186
voice:
   pitch of, 153
   volume of, 179
von Neumann, John, 185

washing machines, 87
Washington, George, 153

watches, 62, *210*
water, *119*, 159, 168
   in baby food, 21
   chemicals in, 52, 54
   hot, for bathing, 120–21
   injected into chicken breasts, 162,
     164
weight, 124
women, differences between men and,
   see gender differences
World War I, 51, 104
World War II, 93, 126, 139